The Challenger Expedition

The Challenger Expedition

Exploring the Ocean's Depths

ERIKA JONES

CONTENTS

Introduction

Funded by the British government and taking its name from the Royal Navy vessel specially converted for the purpose, the *Challenger* Expedition (1872–76) was the first to explore the deep sea successfully on a global scale.[1] Knowledge of the ocean is more important than ever; the modern era is characterised by climate change, species loss, fisheries depletion, ocean pollution and sea level rise. The history of the *Challenger* Expedition and its scientific report provide insights into the advance of oceanography as a modern scientific discipline, a definitive moment in our understanding of the ocean as a complex ecosystem on which all life on Earth depends.

HMS *Challenger* departed Sheerness on the north Kent coast on 7 December 1872, under the command of Captain George Strong Nares and with a remit to explore the physical, chemical and biological characteristics of the deep sea. In addition to the ship's company of almost 250 sailors, engineers, carpenters, marines and officers, there was a six-person civilian scientific team led by Charles Wyville Thomson, a Scottish naturalist determined to prove that life existed in the deepest parts of the ocean. At the time, many within the scientific community doubted whether anything could survive the enormous pressure, cold and darkness of such depths. Others speculated that the ocean floor was covered in a type of primordial ooze, a substance named *Bathybius haeckelii*, connected to the origins of organic life. While the scientists on board the ship had a very specific brief given to them by the Royal Society, *Challenger*'s experienced navigational officers had other reasons to explore the deep sea: the UK Hydrographic Office had given them clear instructions to survey the depth and nature of the ocean floor to support the laying of submarine telegraph cables, a rapidly expanding communications network vital to the commercial and military interests of the British Empire.

Over three and a half years, *Challenger*'s circumnavigation encompassed some 68,890 nautical miles (127,580 km) across the Pacific, Atlantic and Southern Oceans, and traversed the Antarctic Circle. During the voyage, the expedition carried out oceanographic experiments at 504 stations, observing currents, water temperatures, weather and surface ocean conditions.[2]

Along its route, the expedition performed 374 deep-sea soundings, took 255 observations of water temperature, successfully deployed the dredge at 111 stations and completed 129 trawls.[3] Water samples, marine plants and animals, sea-floor deposits and rocks brought up from the deep were carefully preserved on board the ship and sent to Britain for later study.

Between 1880 and 1895, with additional funding from the government, Thomson, and later John Murray, published the *Report on the Scientific Results of the Voyage of H.M.S. Challenger During the Years 1873–76* as a 50-volume series from offices in Edinburgh. Over 75 authors from Britain, Europe and the United States were involved in analysing the specimens and data amassed by *Challenger* and writing the reports. Enriched with information gathered by subsequent voyages, at its completion the *Challenger Report* formed a comprehensive study of Earth's largest and most complicated biosphere. The publication included aspects of marine biology, physics, chemistry and geology, branches of knowledge that would come together to define oceanography as a new scientific field.

The ocean floor beyond the continental shelf was shown not to be a featureless expanse, as many had previously assumed, but instead was characterised by underwater mountain ranges, abyssal trenches and extended plains. The ocean itself consisted of warm and cold water zones, a finding that added to the understanding of ocean currents and the distribution of marine life. Although *Challenger*'s chemist John Young Buchanan debunked the theory of primordial ooze, the investigation of the sea floor led to Murray's discovery of extra-terrestrial particles ('cosmic spherules') in deep-sea sediments. Almost 5,000 new species were found and described, proving that life did exist in the deepest parts of the world's oceans.

There were many difficulties to overcome in conducting the first global study of the ocean, from the collection of hundreds of thousands of specimens to printing and illustrating the reports. It took more than scientific effort and instruments to resolve these issues. Part of *Challenger*'s untold story is how oceanography emerged during a period that revolutionised the movement of people, ideas and commerce around the globe, and how the expedition's results were shaped by Britain's vast empire. Rather than focusing on the better-known characters linked with *Challenger* such as Thomson, Murray and Nares, this book offers a shift in perspective.

The first chapter considers earlier attempts to study the deep sea that influenced and inspired the 1872–76 British voyage. The subsequent chapters delve into the fascinating stories of six objects: HMS *Challenger* itself, the Baillie sounding machine, a mollusc found near Kerguelen Island in the Southern Ocean, a photographic album made by one of *Challenger*'s officers, a deep-sea starfish and the expedition's most far-reaching scientific report, the *Report on Deep-Sea Deposits*. Each of these takes the narrative in new and sometimes unexpected directions, bringing to light the often-hidden work, technologies and people involved in furthering our modern understanding of the ocean.

Tracing the story of the *Challenger* Expedition from multiple perspectives emphasises the enormity of the venture and enables links to be drawn between elements that are commonly overlooked. In Britain, the history of science and exploration is embedded within the country's legacy of empire. Following the travels of these objects, the project's geography includes places not often associated with oceanography, such as Royal Navy bases, transcontinental railways, natural history museums and printers' workshops. In addition to the work of the scientists on board, the *Challenger* Expedition required the labour of hundreds of dockyard workers, coalers, sailors and instrument makers to accomplish its aims. Indigenous Peoples, whalers and naturalists shared their knowledge of harbours, bays, marine animals and features on land. It is notable that the *Challenger Report* was authored by dozens of scientists from several countries, representing a wide range of expertise and skills. Although the voyage of HMS *Challenger* remains a central focus, it becomes clear that the first expansive study of the deep sea was not accomplished by one ship or nation alone.

CHAPTER 1

The Nineteenth-Century Drive to Explore the Deep Sea

> The sea covers nearly three-fourths of the surface of the earth, and, until within the last few years, very little was known with anything like certainty about its depths, whether in their physical or their biological relations. The popular notion was, that after arriving at a certain depth the conditions became so peculiar, so entirely different from those of any portion of the earth to which we have access, as to preclude any other idea than that of a waste of utter darkness, subjected to such stupendous pressure as to make life of any kind impossible.
>
> Charles Wyville Thomson[1]

Powerful interests converged in the nineteenth century in a drive to explore the deep sea. By the late 1860s, intrigued by expeditions that had discovered signs of life at considerable depths, leaders of the British scientific establishment such as William Benjamin Carpenter, Vice-President of the Royal Society, viewed the ocean as one of Earth's most significant unknown environments. Scientific curiosity played a role, of course, but commercial and national motivations also encouraged a greater understanding of the ocean environment, especially knowledge of global currents, fisheries and the topography of the sea floor. Of prime importance was the British government's desire to expand a growing network of undersea telegraph cables that connected London to the administrative, trade and military operations of the British Empire and beyond.

The British Quest to the Poles

In the early nineteenth century, explorers attempted bold ventures into the polar regions. While primarily regarded by the public as heroic quests, these voyages supported an array of scientific investigations, including magnetic observations, meteorology and the collection of materials from the deep sea. In Britain, two influential organisations — the Royal Society, the nation's most esteemed scientific academy, and the Admiralty, the department in command of the Royal Navy — collaborated to launch a series of daring (and sometimes ill-fated) expeditions to find the North-West Passage, a northern sea route that connects the Atlantic Ocean to the Pacific Ocean through a group of Canadian islands known as the Arctic Archipelago. Since the fifteenth century, European explorers had hoped this route could significantly reduce the sailing distance from Europe to Asia, a potential boon for merchant ships and navies, but persistent sea ice and dangerous conditions had stymied all attempts.

Reviving the search for the North-West Passage in 1817, the Admiralty selected Captain John Ross, a 40-year-old Scotsman who was a distinguished veteran of the Napoleonic Wars, to lead an expedition to explore Baffin Bay, a point off Greenland's north-west coast (image 1). Whaling ships had reported a breaking up of ice in the region, news that rekindled notions of a navigable route around the north-east of North America and through the Bering Strait. Leaving London in April 1818 in HMS *Isabella*, Ross was joined by William Edward Parry, commanding HMS *Alexander*. Both ships were loaded with thousands of pounds of beef, bread and raisins, enough supplies to survive a long voyage. Also on board *Isabella* was John Ross's 18-year-old nephew, James Clark Ross. Although young, he had already served six years in the navy under his uncle and had participated in

1 'A Bear Plunging into the Sea', illustration from a drawing made during John Ross's first Arctic Expedition in 1818.

naval actions during the war. The renown gained by discovering the North-West Passage would have been an exciting prospect for the Rosses and the entire ship's company. Following a forceful northern current that whalers had observed along the Canadian coast, by early June the two ships were trapped in a semi-frozen strait.[2] Pressing forward, the crews dragged the *Isabella* and *Alexander* through the slush, eventually making it to Baffin Bay in the middle of August.[3]

While exploring the bay and its various inlets, the expedition conducted scientific observations, including measuring the temperature of the water and gathering samples from the ocean floor. John Ross deployed a device that he invented and described as a 'Deep-Sea Clamm'. Made by *Isabella*'s armourer, the hollow cast-iron container procured substances from great depths and brought them back to the surface for examination. With a hinged opening held open by a bolt at one end, the device measured 18 inches (45.7 cm) long by 5 inches (12.7 cm) wide. After being dropped from the ship and on reaching the sea floor the hinges snapped shut, trapping the contents inside (image 2). In order not to lose the instrument during its long descent and subsequent retrieval, it was fixed to the strongest rope available: 2½ inches (6.4 cm) in diameter and made of the best hemp.[4] Using the rope, the depth of the water was measured in fathoms, one fathom being 6 feet (1.8 m). (The word 'fathom' comes from the Old English *faedm* or *faethm*, meaning the distance between a man's outstretched arms.)

On 1 September 1818, John Ross's Deep-Sea Clamm sank to a depth of 1,000 fathoms (1,828 m); it took 27 minutes to descend and an hour for all hands to raise it to the surface again. Ross wrote that the material captured from the sea floor consisted of 'soft mud, in which there were worms, and, entangled on the sounding-line, at the depth of 800 fathoms (1,463 m), was found a beautiful *Caput Medusae*', later identified

2 A half-size model of a Deep-Sea Clamm of the type invented in 1818 and used by Captain John Ross on board HMS *Isabella* in Baffin Bay that year.

as a type of starfish with many arms (image 3).[5] The remarkable creature remained intact during its trip to the surface, making it the earliest recorded instance of a live animal brought up from such prodigious depths.[6]

When the expedition returned to England in November 1818, however, there was little cause for celebration. The North-West Passage remained elusive and the voyage was clouded in controversy. During the expedition's investigation of Baffin Bay on 30 August 1818, *Isabella* entered Lancaster Sound, a channel between two islands. After days of sailing through fog and gales, John Ross saw a mountain range on the distant horizon. Believing that the way was blocked, he decided to turn back, causing consternation among his junior officers who wished to explore further. Defending his decision, Ross later wrote: 'it appears perfectly certain that the land is here continuous, and that there is no opening at the northernmost part of Baffin's Bay'.[7] The Admiralty was not as convinced and the next year Parry was given command of an expedition to return to Baffin Bay. Reaching Lancaster Sound, Parry continued west and discovered and named Prince Regent Inlet before being blocked by ice. After surviving a winter trapped in the frozen sea, Parry returned to England in October 1820 with the jubilant news of the expedition's achievement.

To restore his reputation, Ross desired to search for the North-West Passage again. However, the government refused to back his plans. Aid came from a wealthy businessman interested in Arctic exploration, Felix Booth, who decided to finance privately the British North-West Passage Expedition, 1829—33. Over the previous decade Booth had expanded his family's distillery company and amassed a personal fortune; it was gin profits that paid for Ross's use of a reinforced steamer, auspiciously named *Victory*, and supplies to last several years.

On 23 May 1829, with his nephew as second-in-command, two additional officers and 19 men, Ross left the Thames in London and headed for the Arctic. By September, they had passed through Lancaster Sound and into Prince Regent Inlet and proceeded south. When the ship became engulfed by ice, the crew spent four winters in the Arctic. As each spring and summer came, attempts were made to break free, but progress was slow. A group of Netsilik Inuit provided much-needed food and information, without which Ross and his men may well have struggled to survive. Although they

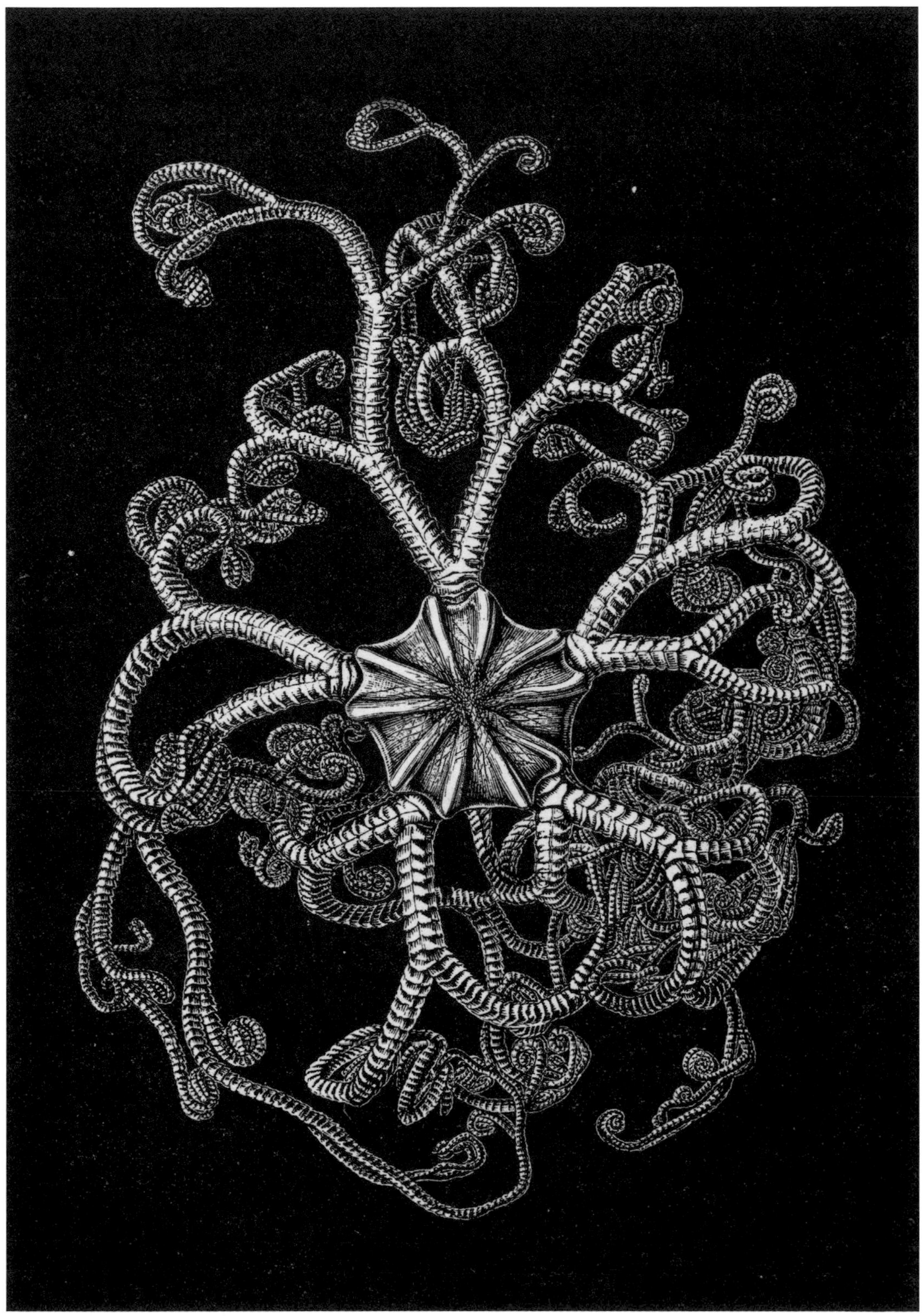

3 An illustration of *Asterophyton linckii*, 'Medusa's Head Basket Star', a species found in the deep Atlantic by Captain John Ross in 1818.

4 *Captain Ross and the crew of the 'Victory' saved by the 'Isabella' of Hull*, by Edward Francis Finden, 1833.

faced hardships daily, the expedition continued to make scientific and meteorological observations, including explorations overland by sledge. Using a dip circle (an instrument for measuring the inclination, or dip, of Earth's magnetic field) and the star Polaris to navigate, James Clark Ross led a party that reached the North Magnetic Pole on 1 June 1831.[8]

By January 1832, it became clear that *Victory* would remain stuck in the ice and John Ross made the decision to abandon ship. That spring, the crew journeyed 300 miles (483 km) north on foot to Fury Beach at Lancaster Sound, where they attempted to escape by boats that Parry (along with James Clark Ross) had abandoned after the loss of HMS *Fury* in 1825. Finding the way again blocked by ice, they returned to Fury Beach and spent yet another winter in the Arctic. Incredibly, the group was spotted by a passing whaling ship and was rescued that August. In a twist of fate, it was *Isabella*, the same vessel that John Ross had commanded during his first foray into the Arctic (image 4). On their return to England in 1833, he and his nephew were greeted with a hero's welcome (image 5). Adding to their accomplishments, the expedition brought back marine specimens that provided rare glimpses of life in the Arctic seas (image 6).[9]

5 *Commander James Clark Ross, Discoverer of the North Pole*, by John Robert Wildman, 1834. This portrayal marks Ross's return from the 1829–33 Arctic expedition. The Pole Star shines in the sky in the painting's top right and in the lower right corner is a magnetic dip circle, both of which aided his excursion to the North Magnetic Pole.

During this period, the UK Hydrographic Office (established in 1795 as the Hydrographical Department of the Admiralty to compile charts and information for improving navigation) continued to support expeditions that furthered a more scientific understanding of the ocean. Leading the department from 1829 to 1846 as Hydrographer of the Navy, Rear-Admiral Sir Francis Beaufort commissioned a fleet of survey vessels to acquire information on coastlines, winds, currents and tides. Throughout his career, Beaufort strove to improve navigation by systematically observing and recording ocean phenomena. One of his early accomplishments included developing a simple visual system to help mariners measure wind intensity based on sea conditions in 1805. The Beaufort wind force scale is still used in marine weather forecasts today.

Beaufort was a Fellow of the Royal Society and had a close association with the British scientific community, a relationship that fostered the Admiralty's support of a British naval expedition to the Antarctic in 1839 to carry out magnetic measurements. Having worked with the British Magnetic Survey from 1834 onwards, and with his considerable polar experience, James Clark Ross was chosen to lead the expedition. He commanded HMS *Erebus* and his close friend Captain Francis Crozier was in charge of HMS *Terror*. During the voyage, the expedition charted much of the Antarctic coastline. Following the wishes of the British Association for the Advancement of Science (BAAS), James Clark Ross measured Earth's magnetic field and established the position of the South Magnetic Pole. While charting part of the eastern coast of Antarctica, the expedition discovered an enormous 'ice barrier' more than 370 miles (600 km) long and between 16 and 55 yards (15 and 50 m) high above the sea,

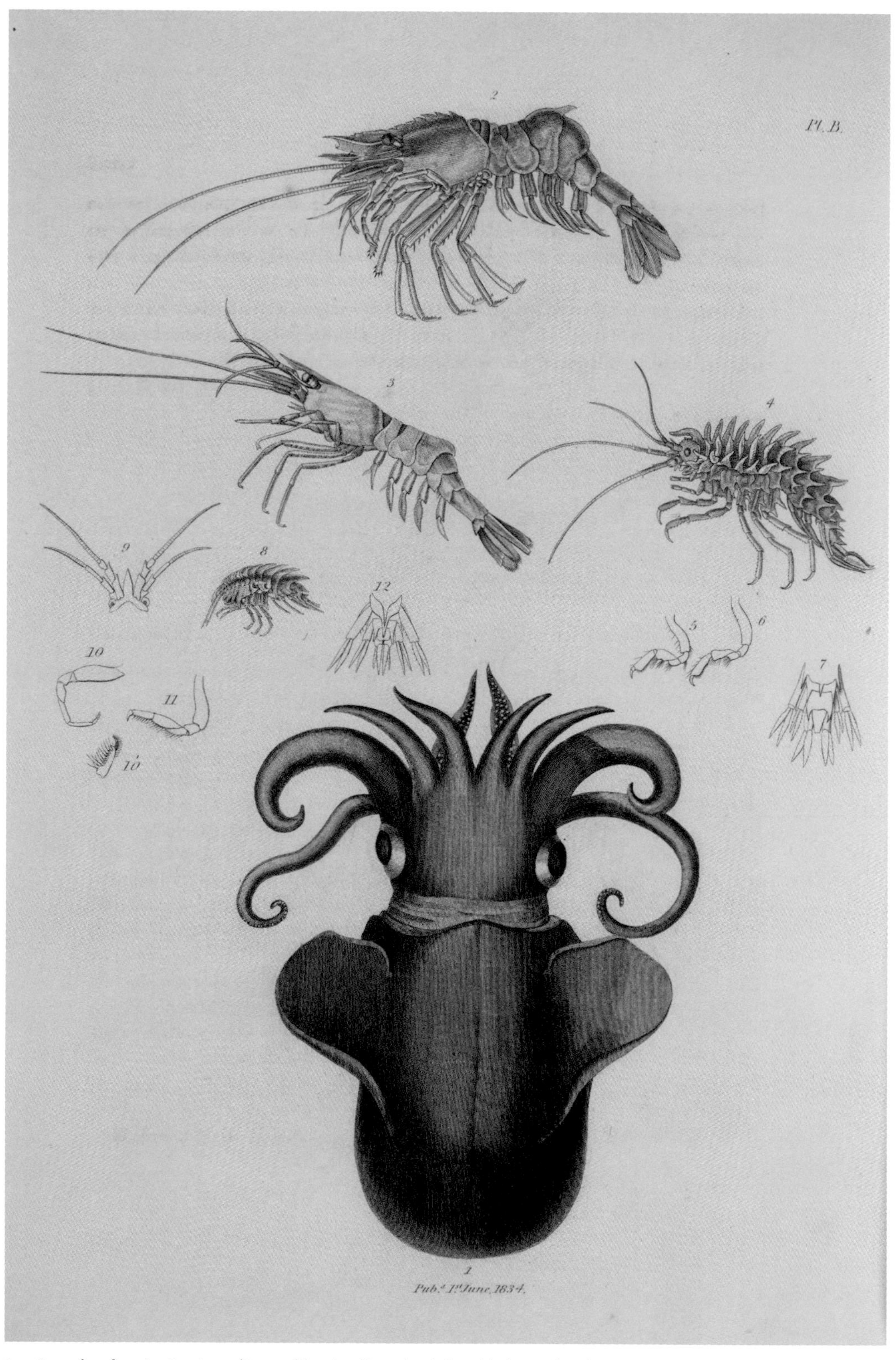

6 Examples of marine Crustacea (Decapoda) and Mollusca (Cephalopoda) observed on the Ross Arctic expedition, 1829–33.

now named the Ross Ice Shelf, which prevented ships from sailing further south.[10]

In addition to magnetic observations, the British Antarctic Expedition conducted what could be considered oceanographic research today. There was no separate scientific team, so the naval officers and crew carried out investigative work such as recording meteorological conditions, tides and currents. The officers also took soundings (measurements of the depth of a body of water), recorded ocean temperatures and collected water samples. The young naturalist Joseph Dalton Hooker, who later went on to become a distinguished botanist and the director of the Royal Botanic Gardens, Kew, was employed as assistant surgeon in *Erebus*. In addition to gathering plants on land, he deployed tow nets from the ship to capture sea life and algae.

During their explorations in the southern polar regions, the crew discovered signs of life deep below the ocean surface. In *Erebus*, James Clark Ross deployed a Deep-Sea Clamm and dredge (an instrument used to bring up objects or material from the seabed), successfully obtaining living creatures in sea-floor deposits taken from depths of 300 fathoms (548 m) and more (image 7).[11] In his account of the expedition, he expressed his excitement at discovering coral from a depth of 270 fathoms (493 m) on 19 January 1841 in an area of the Southern Ocean now known as the Ross Sea:

> Becalmed for two or three hours after noon, the dredge was put over in two hundred and seventy fathoms water, and after trailing on the ground for some time was hauled in. ... The most remarkable circumstance was drawing up from so great a depth beautiful specimens of living coral, which naturalists and geologists have hitherto concurred in believing unable to work beyond the pressure of a few fathoms below the surface. Corallines, Flustræ, and a variety of marine invertebrate animals, also came up in the net, showing an abundance and great variety of animal life.[12]

Although 'contrary to the general belief of naturalists', James Clark Ross had 'no doubt that from however great a depth we may be enabled to bring up the mud and stones of the bed of the ocean we shall find them teeming with animal life', even in the extreme cold of the polar regions.[13]

7 *Beaufort Island and Mount Erebus. Discovered 28 Jan 1841*, by John Edward Davis, 1841.

Naturalists at Sea

National expeditions received public attention and fanfare, but routine naval survey cruises also provided opportunities to study marine environments. In 1831 Captain Robert FitzRoy, a 26-year-old aristocrat and naval officer interested in science, desired a gentleman naturalist to accompany him during HMS *Beagle*'s voyage around the world. This request was perhaps for his own well-being. *Beagle*'s previous captain, Pringle Stokes, had met a tragic fate: after two years of dangerous work surveying the rocky inlets of the Strait of Magellan, Stokes became isolated, severely depressed and ultimately shot himself in the head. While preparing for his forthcoming mission, FitzRoy asked Beaufort to find a suitable companion who could make good use of the expedition's opportunities for studying geology, plants and animals on shore and to dine with him as an equal.[14] Hearing of the appointment, John Henslow, Professor of Botany at the University of Cambridge, wrote to his former student, 22-year-old Charles Darwin, 'there never was a finer chance for a man of zeal & spirit ... Don't put on any modest doubts or fears about your disqualifications for I assure you I think you are the very man they are in search of.'[15] Beaufort gave his approval and, after FitzRoy met and dined with him, it was decided that Darwin would go.[16]

Later the same year, on 27 December, *Beagle* sailed from Plymouth for South America. Over the course of almost five years, FitzRoy and the crew carried out an extensive hydrographic survey along the east and west coasts of South America to the Galápagos Islands, before sailing west across the Pacific to visit Tahiti, New Zealand, southern Australia, Mauritius in the Indian Ocean and the Cape of Good Hope, South Africa. Finally, they returned to England in October 1836. During the voyage, Darwin developed his theory of coral reef formation. The Scottish geologist Charles Lyell had argued that corals formed on the top of volcanoes rising from the ocean floor. From observing barrier reefs and atolls, Darwin formulated an alternative explanation. If the land was being pushed up, the coral would become exposed above sea level and die, a phenomenon that he could not envision. Although it seemed miraculous, he supposed the opposite process must be true: corals are continually building upwards towards sunlight and reef formations occur as land subsides into the sea. Referring to islands in the Pacific, including Tahiti and Eimeo (now called Moorea), Darwin declared, 'We must look at a Lagoon [Island] as a monument raised by myriads of tiny architects, to mark the spot where a former land lies in the depths of the ocean.'[17]

In 1842 Darwin published his first scientific book, *The Structure and Distribution of Coral Reefs*, describing his hypothesis (image 8). Although Lyell quickly accepted Darwin's new theory, the mechanics of coral formation were debated for many years to come. Following nuclear tests by the US government in the 1950s, cores taken from Bikini Atoll were examined and dated and Darwin was finally proven correct. The island's coral reef had been growing upwards for the

8 'Coral island and circling coral reef creating a lagoon', included in Charles Darwin's 1842 book on coral reefs. The realisation of the long timescale involved in the creation of islands and reefs contributed to Darwin's theory of evolution.

9 A lively cartoon drawn by Edward Forbes in 1859 showing his use of a dredge, a key tool for early marine naturalists to collect specimens from the ocean floor.

last 30 million years as the volcano below sank slowly into the sea.

Like Darwin, British naturalist Edward Forbes depended on his inherited wealth and social position to support his interest in natural history. In the 1830s Forbes was one of the early adopters of the 'dredge', a modified and weighted fishing net used to collect specimens from the ocean floor (image 9). He worked closely with the BAAS's Dredging Committee, a group that provided a like-minded community as well as grants for publication and illustration. Dredging in relatively shallow waters near shore could be profitably accomplished from a yacht and was often a sociable gathering that included men and women. Known as a likeable and genial character, Forbes penned a whimsical song about the joys of dredging:

> Hurrah for the dredge, with its iron edge,
> And its mystical triangle,
> And its hided net with meshes set,
> Odd fishes to entangle!
> The ship may rove through the waves above,
> Mid scenes exciting wonder;
> But braver sights the dredge delights
> As it roveth the waters under!
> (Chorus)
> Then a dredging we will go, wise boys!
> Then a dredging we will go![18]

Around this time, women participated in the sailing, collected and observed marine fauna and flora, and were active members of natural history societies and clubs.[19] As the discipline of marine zoology became more professionalised, however, amateurs and women were increasingly excluded from research and work settings such as naval vessels.[20]

Forbes benefited from his friendship with Captain Thomas Graves, who convinced the Admiralty to appoint him as honorary naturalist in HMS *Beacon* during its 1841–42 survey in the Aegean Sea. During the voyage, Lieutenant Thomas Abel Spratt and the naval crew conducted more than 100 dredgings along the southern coasts of Greece and Anatolia, up to 230 fathoms (420 m) deep. Forbes recorded the plants and animals found, and sought out patterns of distribution. He identified four different 'zones' of life. They ranged from the littoral, the space between the tide marks on shore, to the deepest waters inhabited by corals. He claimed the lowest region extended from about 50 fathoms (90 m) to an unknown lower limit. In 1843 Forbes presented his findings to the BAAS; generalising his observations beyond the Mediterranean Sea, he speculated that animal life was probably not present below the region of 300 fathoms (548 m). His writing on the subject influenced the next two decades of ocean science and how intellectuals regarded the deep sea: 'As we descend deeper and deeper in this region its inhabitants become more and more modified, and fewer and fewer, indicating our approach towards an abyss where life is either extinguished or exhibits but a few sparks to mark its lingering presence.'[21]

Despite the evidence of deep-sea life previously found by polar expeditions, what became known as the azoic or 'lifeless-depths' hypothesis captured the imagination of many European and American scientists. Part of the debate derived from the source of evidence. Forbes's research garnered more respect in academic circles than the previous observations of naval officers. In addition, due to the extreme pressure, darkness and cold found in the deep ocean, few naturalists challenged the 'lifeless' hypothesis. It would take more evidence, as well as improved deep-sea dredging instruments, to prove definitively that animals lived below 300 fathoms. Yet, Forbes himself acknowledged that there was still much to be discovered. Shortly before his death, he wrote, 'it is in the exploration of this vast deep-sea region that the finest field for submarine discovery yet remains'.[22]

Alongside Britain, other maritime nations sponsored expeditions in the 1830s and 1840s to explore the ocean. In 1836 King Louis Philippe I of France (1830–48), launched notable excursions that combined diplomatic and imperial aims with scientific investigations. Commanded by naval officer Auguste-Nicolas Vaillant, the former troopship *Bonite* voyaged for 21 months around the globe from 1836 to 1837. The ship delivered French diplomatic and consular representatives to Chile, Peru and the Philippines, and visited trade ports and religious missions in South America and Hawai'i. Scientific enquiry also played a significant role. The naturalists Joseph Eydoux and Louis Souleyet, along with two watercolour artists, accompanied the voyage. Some 3,500 natural history specimens were brought back to the National Museum of Natural History in Paris, and drawings of fish, crustaceans, molluscs and terrestrial animals made during the voyage formed the basis of the expedition's richly illustrated zoological report (images 10 and 11).[23]

The *Bonite* also made strides in physical oceanography. Officers attempted to record the temperature of the water at a variety of depths, but the cylinders that encased the thermometers were not strong enough to protect them from damage from the increased pressure in deeper water. They had more success using a water collecting bottle devised by Jean-Baptiste Biot, a renowned French physicist and astronomer.[24] Made of thick glass and brass, the apparatus was lowered into the water on a weighted line (image 12). Pulling on another rope attached to a siphon would draw water into the instrument's tube, where it was held until it was brought back to the surface. The seawater was then examined for its salinity and

10 Drawing of a mollusc discovered during the voyage of the *Bonite* (1836–37), attributed to 'Borromée'.

11 Hand-coloured illustration of a fish observed during the voyage of the *Bonite*, by Paul Louis Oudart.

chemical composition, two aspects that oceanographers still research today. Although most seawater has a salinity of between 31 g/kg and 38 g/kg (or roughly 3.1 per cent), it is not uniform throughout the world. Seawater can be substantially less saline where mixing occurs with freshwater run-off from river mouths or near melting glaciers, for example. The salinity also affects ocean currents. Warm water can hold more salt than cold water; as warm, salty water gains density, it begins to sink through colder water, creating a convection current.

Ocean physics was further advanced by the French *Vénus* Expedition (1836–39) under Captain Abel Aubert du Petit-Thouars. Leaving in 1836 from Brest, a port city in Brittany, the expedition sailed around the world over two and a half years. Although Petit-Thouars's primary mission was to protect French traders, missionaries and diplomatic interests in the Pacific, the officers followed a scientific programme at sea. The crew collected reams of data, including hourly meteorological and sea-surface temperature

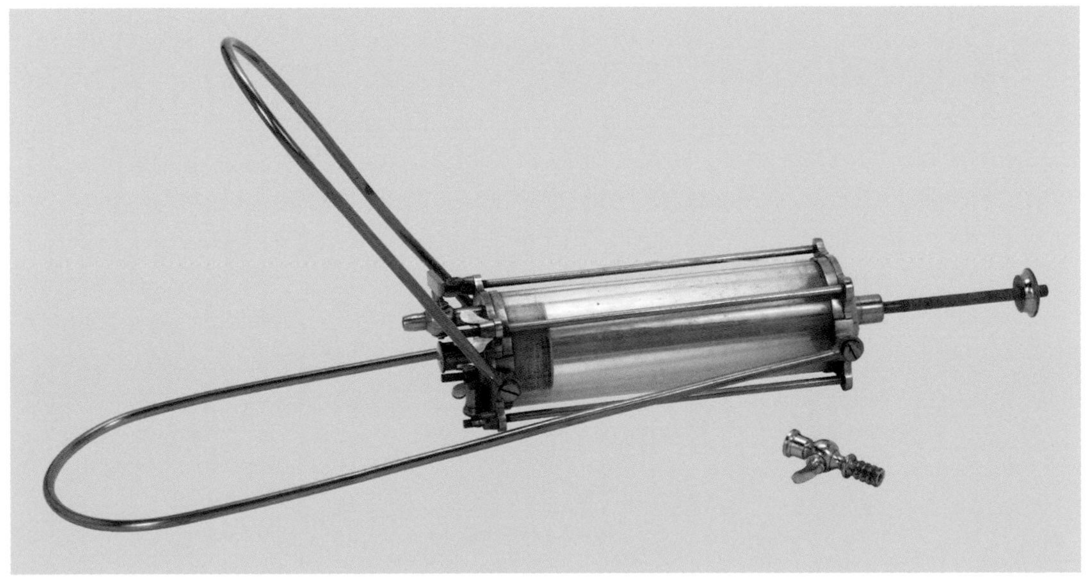

12 Water collecting bottle of the type used on the *Bonite* to sample water from great depths.

readings, observations on surface currents and ocean temperatures at different depths. As a result, *Vénus*'s findings provided an emerging picture of how the ocean circulates. In his report after the voyage, French hydrographer-engineer, physicist and cartographer Louis Urbain Dortet de Tessan asserted that cold water from the poles generally flows towards the equator along the western coasts of continents, and warm water flows from the equator towards the poles on the eastern side of continents. While this is only one factor in the complex nature of ocean currents, his general observation later proved correct. From temperature readings at different depths, Tessan also deduced that below the surface current, there could exist a deeper current with a distinctive temperature and direction.[25]

In the middle decades of the nineteenth century, the United States contributed considerable momentum to ocean exploration. As with France and Britain, the US government launched voyages that furthered its imperial ambitions during this era. The United States Exploring Expedition (USEE) (1838–42) was a bold naval operation that projected American military strength throughout the Pacific. One of the American navy's most talented nautical surveyors, Lieutenant Charles Wilkes, was given command. Wilkes led the navy's Department of Charts and Instruments, an organisation that later became the Naval Observatory and Hydrographic Office. Showcasing the growing naval fleet, the convoy consisted of six sailing vessels and 346 men, including a team of nine civilian scientists and artists. Even with its impressive size, the expedition encountered troubles, especially given the temperament of its commander. Wilkes gave scientists little authority on board and was later court-martialled for the excessive punishment of his sailors. After his nephew Midshipman Henry Wilkes was killed by Fijians, he ordered a massacre that killed 87 people on the island of Malolo.[26] Competing for trading outposts and colonies in the Pacific, imperial nations such as Britain, France and Germany also inflicted violence upon the Indigenous Peoples of Oceania during this period.[27]

After losing two ships and 28 men, the Exploring Expedition finished its circumnavigation and brought back a great deal of material, both ethnographic and zoological. While most of the 50,000 specimens were terrestrial, geologist James Dana also collected 400 species of coral and 1,000 species of Crustacea. In 1858 the collection eventually found a home in America's first national museum, the

Smithsonian Institution, increasing the organisation's prestige and inspiring a new generation of marine researchers. Despite his actions during the voyage, Wilkes was advanced to the rank of commander in 1843. From 1844 to 1861, he championed and received government funds to publish an expedition report, which included 15 scientific reports alongside his five-volume narrative of the voyage. Although the other reports varied in quality, Dana's study of Crustacea, in which more than 500 new species of lobster, crab, shrimp and barnacle were identified, was a milestone for the field.[28]

Following the United States' conquest of California and the territory's annexation in 1848, the expanding nation looked to further its influence westwards. The North Pacific Exploring Expedition (1853–56) was organised by the US Navy Department and represented the first major effort by any nation to study the ocean, rather than concentrating heavily on coastlines, islands and ports. Consisting of five ships, its mission was to survey 'for naval and commercial purposes such parts of the Bering Straits, the North Pacific Ocean, and the China Seas, as frequented by American whaleships, and by trading vessels in their routes between the United States and China'.[29] Captain Cadwalader Ringgold, who had been third-in-command on Wilkes's voyage, was placed in charge of the expedition, which departed from Norfolk, Virginia, in June 1853 and explored the coasts of China and Japan, Madeira Island, California and Tahiti before returning via the Cape of Good Hope in 1856. Extensive natural history collections were made, mostly by chief zoologist William Stimpson, one of the country's foremost naturalists who worked with a dredge. He discovered and named hundreds of marine invertebrates during the voyage.[30]

As part of the expedition's work, Midshipman John Brooke conducted the first confirmed deep-sea soundings in the Pacific.[31] He deployed a device he invented, which became known as the Brooke's sounder, and brought up a small amount of material from the seabed (image 13). This meant that Stimpson, aided by a microscope on board, was able to study fresh sediments, something that had not been possible before. Despite these accomplishments, when the voyage returned, the official results were never published. The failure to do so was partly caused by Ringgold suffering a mental breakdown during the voyage, and the disruption caused by the American Civil War that began in 1861. However, the expedition had a lasting

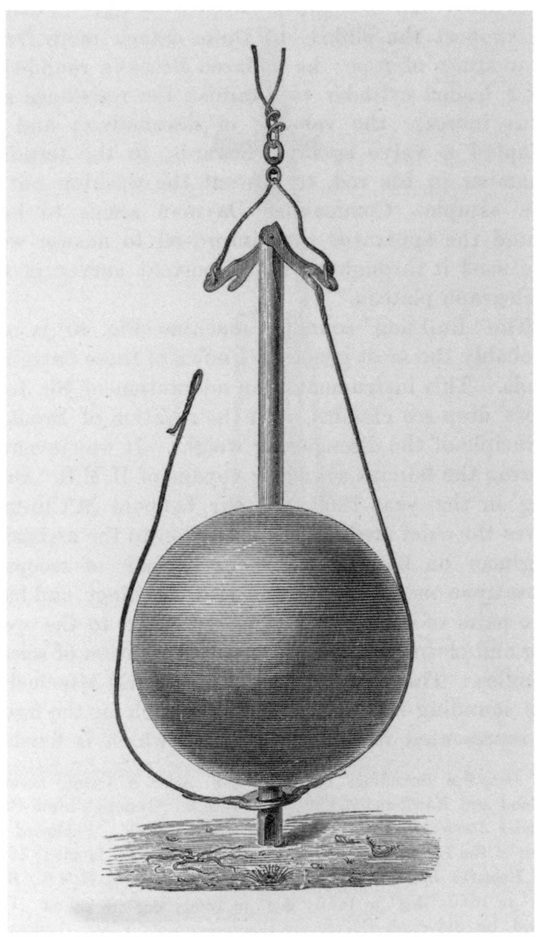

13 'Brooke's Deep-Sea Sounding Apparatus', as used on the US navy's North Pacific Exploring Expedition, 1853–56.

impact on the emerging field of oceanography and prompted new questions about the deep sea. Crucially, Stimpson shared valuable knowledge about dredging on a naval ship with British and American naturalists, aiding future expeditions.[32]

Expanding the Worldwide Telegraph System

As naturalists and hydrographers became more curious about the possibility of life in the ocean abyss, a radical new communications system, the electrical telegraph, further propelled ocean exploration. By 1852, national telegraph systems operated in the US, the UK, Prussia, Austria, Canada and France. Messages that once took days or weeks to deliver by overland post or carry by ship could now be sent electronically within minutes. Laying submarine telegraph cables had the potential to connect existing continental networks into a worldwide system of electronic communication. But engineers knew little of the ocean environment in which undersea wires were intended to operate and even less about the sea floor upon which a cable would sit.

On land, the electrical telegraph followed the expansion of railways. Telegraph poles and wires sprung up alongside tracks, and stations for sending and receiving telegrams were installed in post offices. Laying undersea cables, posed a greater technological challenge. Part of the problem was that wires had to be well insulated to prevent the electric current leaking into the water. In 1842 American inventor Samuel Morse used tarred hemp and India rubber to encase a cable that he then submerged in New York Harbour, proving that running electric cables underwater was indeed possible.

Later the same year, William Montgomerie, a Scottish surgeon in the British East India Company's service, introduced gutta-percha, a natural plastic formed from the sap of *Palaquium gutta*, a tree native to the Malaysian Peninsula. Using the new material, in 1847 Prussian army officer William Siemens laid an underwater cable across the Rhine River, connecting the German cities of Deutz and Cologne. This achievement inspired ideas for more ambitious projects that would begin to link islands with continents.

Seeing the telegraph's potential, inventor and entrepreneur John Watkins Brett and his younger brother Jacob, sons of a Bristol cabinetmaker, registered a business they called the General Oceanic Telegraphic Company. In July 1845 they submitted a patent to the British government, which declared: 'By means of this telegraph any communication may be instantly transmitted from London or any other place, and delivered in a printed form, almost at the same instant of time, at the most distant parts of the United Kingdom or of the Colonies.'[33] Perhaps these claims seemed too far-fetched for government administrators as their bid for funding was rejected. In 1847 the brothers went to France where they had more success; Louis Napoléon, President of the French Republic, granted them a ten-year contract to lay cables from England to France. On 28 August 1850 the two brothers, at their own cost, laid the first submarine telegraph cable in the open ocean across the turbulent waters of the English Channel. Made of a single strand of copper wire coated with gutta-percha, it transmitted messages for a few hours. The system was fragile, though, and soon the line was caught by a fisherman's anchor and broken.[34]

Undeterred, the Brett brothers laid another cable across the Channel the following year, this time with a more protective covering. The cable was sheathed in iron wire, wrapped in jute and coated with pitch. The stronger line worked well and successfully connected the British and French telegraph networks. The production of more undersea cables quickly followed and by 1852 Britain was linked by undersea telegraph to Ireland, Belgium and the Netherlands. It was clear that submarine telegraphy had the potential to cross even greater distances.

The growth of the submarine telegraph network presented many advantages to the British government, as the forward-looking brothers recognised in 1845. By transmitting messages between merchants, traders and private individuals, the technology strengthened London's control over its colonies and supported British commerce in distant locales across the globe. An early example of the telegraph as a military tool was its use between 1853 and 1856 in the Crimean War, one of the first conflicts to employ modern technologies such as railways and heavy artillery. During the war, the UK, France and the Ottoman Empire fought Russian forces for two and a half years. In April 1855, the laying of a 340-mile (547 km) submarine cable across the Black Sea, between the seaside towns of Balaklava (Crimea) and Varna (Bulgaria), enabled officials in London and Paris to communicate with their commanders in Crimea within 24 hours. For the generals, increasingly plagued by political interference, this proved a mixed blessing. General James Simpson, commander of British troops in Crimea, complained that 'the confounded telegraph has ruined everything'.[35] With mismanagement and disease rampant on both sides, Russia relented and accepted peace terms in 1856.

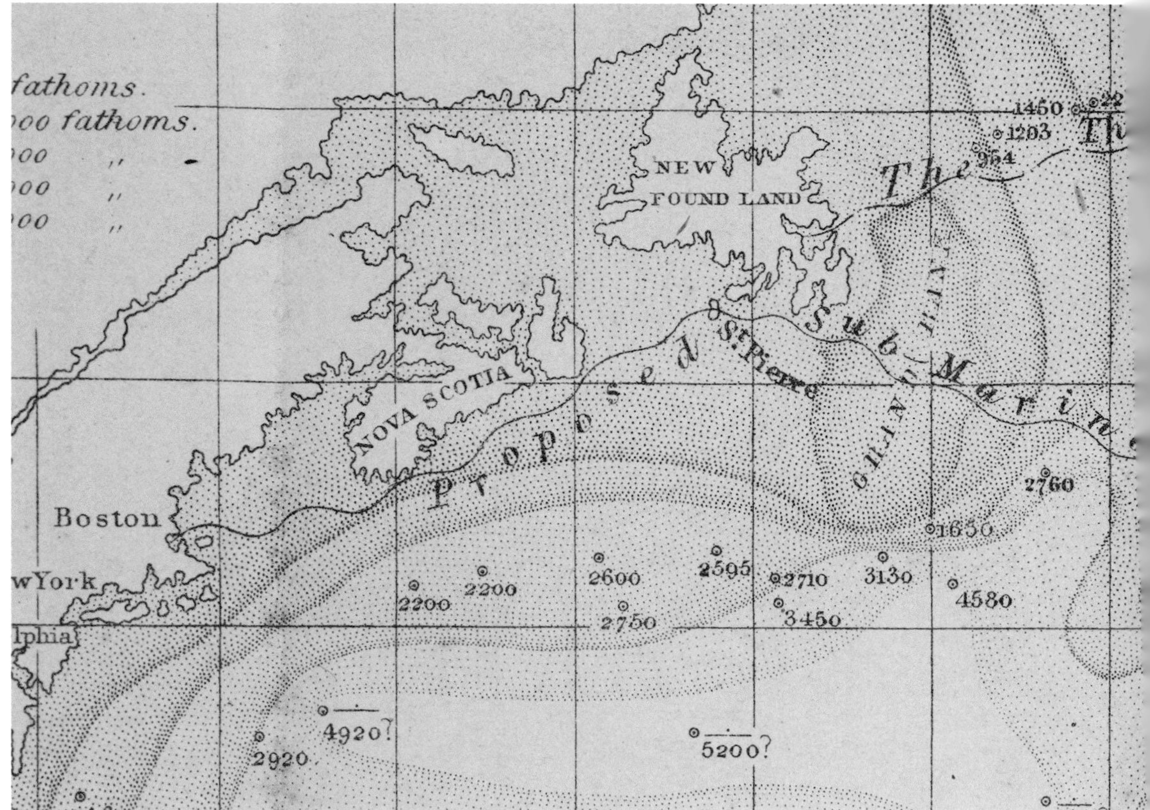

14 Chart showing the 'Telegraphic Plateau', by Matthew Fontaine Maury, published in 1861.

The advantages of rapid telegraph communications were not lost on the United States. In 1854, using Brooke's sounding device and information gathered from several voyages, American naval officer and pioneer hydrographer Matthew Fontaine Maury, Superintendent of the United States Naval Observatory and Head of the Depot of Charts and Instruments, issued the first chart indicating the depth of water in the North Atlantic. The chart showed a relatively flat and shallow section of the bed of the Atlantic that begins around latitude 51° N, which Maury dubbed the 'Telegraphic Plateau' (image 14).[36] Maury's findings convinced American businessman and financier Cyrus West Field that a cable could be laid along thc submerged 'plateau', connecting the British Isles and the Americas via a 1,400-mile (2,300 km) route from Ireland to Newfoundland.

Field's expertise was in business management and he had made his fortune in paper manufacturing. Working with capital investors and engineers, he led a group to form the Atlantic Telegraph Company, which began laying its first cable in 1857. After many trials and

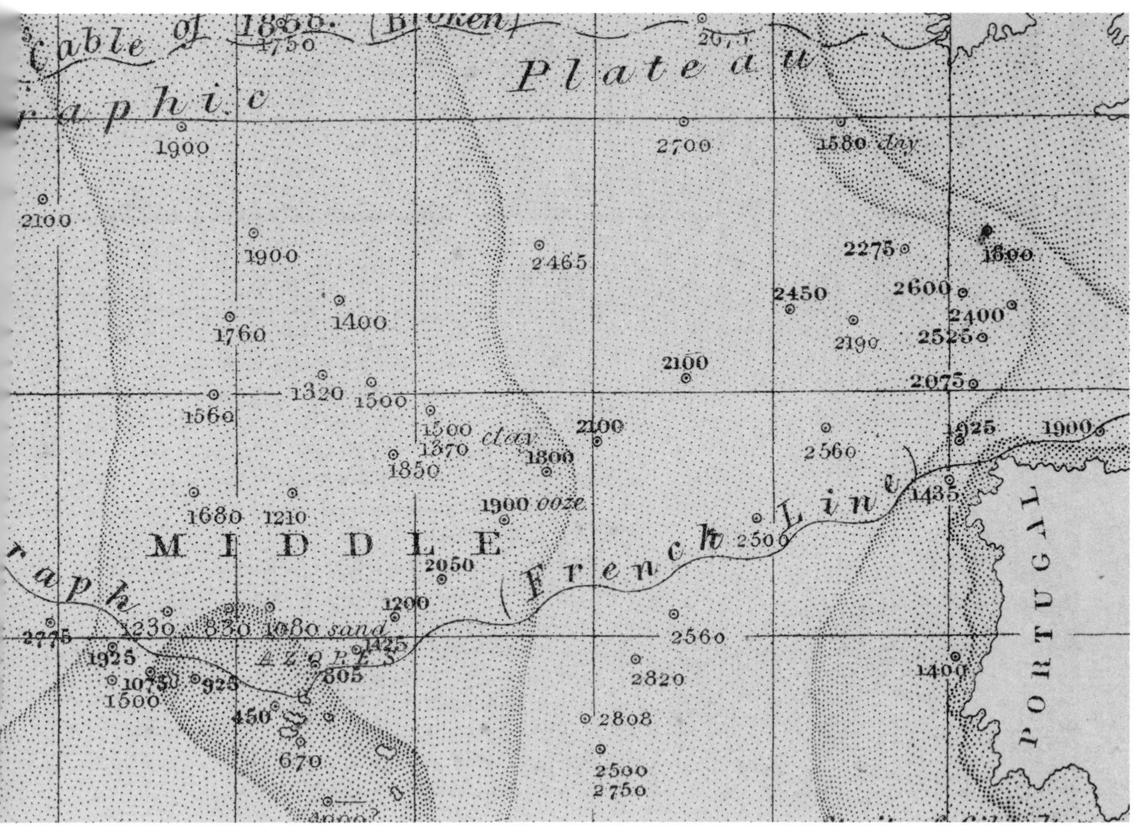

15 Sample length of the 1858 Atlantic Telegraph submarine shore-end cable. Copper and iron wires sheathed in hemp, India rubber and pitch.

setbacks in which cables were broken, lost and repaired, on 29 July 1858 two specially adapted naval ships — the American steam frigate USSF *Niagara* and the British HMS *Agamemnon* — steamed to the North Atlantic, each vessel carrying 1,100 nautical miles (2,037 km) of cable (images 15 and 16). At 52° 59' N, 32° 27' W, the two massive cables were spliced together and the ships began to pay out the line in opposite directions. Travelling east, *Agamemnon* brought one end of the cable to Ireland, and the western-bound *Niagara* carried the other end to Newfoundland.

This time, calm weather and seas aided their work and, a week later, the two ships arrived safely at their respective destinations. On 5 August 1858, Field recorded in his diary the moment when the *Niagara* arrived at Bay Bulls Arm, Newfoundland, and the cable was first activated:

> At 5:15 A.M. the telegraph cable was landed. At 6 A.M. the shore end of the cable was carried into the telegraph house, and a strong current of electricity received through the whole cable from the other side of the Atlantic. Captain Hudson then read prayers, and made some remarks. At 1 P.M. the steamer *Gorgon* fired a royal salute of twenty-one guns.[37]

The occasion was heralded as one of the most spectacular technological achievements of the age. The cable was celebrated in New York City with a parade and fireworks, and Tiffany & Company sold small lengths of the surplus cable as souvenirs.[38]

By September, to great consternation and disbelief, the transatlantic cable had ceased functioning. Field was eager to try again and the British government continued to provide funding. The first cable probably broke down due to its small diameter, deterioration of the insulator and its inability to sustain the applied voltage. A special government committee of scientists and engineers reviewed the failure and declared that a 'careful and detailed survey of the nature and inequalities of the bottom of the sea' was needed before future cable laying took place.[39] Knowing the landscape of the seabed — whether the cable was to pass over flat stretches of soft mud or climb submerged mountains and rocky outcrops — determined the length of cable required. Moreover, without the proper armouring, the entire process was likely to fail once more. In response, British hydrographers turned their attention to charting the ocean floor as a matter of national importance. Engineers also grappled with cable designs and the complicated mechanics of laying and repairing cables situated at the bottom of the sea.

Meanwhile, the American Civil War disrupted work on the other side of the Atlantic. In the 1850s, Maury was an international leader in the organisation of meteorology and ocean science. His efforts resulted in an expansive series of charts that not only shortened routes for mariners but also represented winds, currents, depths, temperatures and even biological information.[40] *The Physical Geography of the Sea*, which he published in 1855, is generally regarded as the first oceanographic textbook.[41] As part of Maury's support of American imperial power, however, he was an outspoken proponent of white supremacy and lobbied aggressively for expanding the institution of slavery from the southern US to South America.[42] At the onset of the Civil War in 1861, he joined the Confederacy and

16 *The 'Niagara', 'Valorous', 'Gordon' & 'Agamemnon' Laying the Cable at Mid-Ocean*, by Sarony, Major & Knapp, 1861.

went to England as a special agent to purchase vessels for raiding Union commerce ships, thus ending his oceanographic career.

More than a decade after its initial beginnings and failure, a permanent undersea transatlantic cable was laid in 1866. Engineers had learned from their experience with undersea cables in the Mediterranean and the Red Sea, with the result that British manufacturing centres produced an improved design. At nearly twice the weight of the 1858 cable, the new line consisted of a core of seven twisted strands of very pure copper covered by layers of waterproof adhesive, gutta-percha, hemp, woven strands of high-tensile steel wire and fibrous yarn. The SS *Great Eastern* was used to lay it in 1865, but the line snapped and was lost in the mid-Atlantic (image 17). After returning to England, further capital was raised for another attempt and the Anglo-American Telegraph Company was formed. Laying commenced on 13 July 1866. Two weeks later, on 27 July, the *Great Eastern* arrived at Heart's Content, Newfoundland, and the cable was put into operation, linking it with a station on Valentia Island, one of Ireland's most westerly points.[43] For the first time, a message could be sent across the Atlantic and a response received during the same day — a true revolution in communication. Over the next few decades, many more undersea cables were laid around the globe, creating the beginnings of a vast network of wires that we know as the internet today.

Search for Life in the Abyss

As a result of raising submarine telegraph cables for repair, new evidence emerged of the existence of life in the deep sea. In October 1860, a cable between the island of Sardinia, Italy, in the Mediterranean, and Annaba (then Bona), Algeria, on the northern coast of Africa, failed, most likely broken by the anchors of coral fishers. Renowned English engineer Henry Charles Fleeming Jenkin supervised the repairs on a ship equipped with specialised machines, wooden buoys, ropes and chains to lift the cable to the surface. Writing to his wife Ann, he described the process of recovering the line: 'What rocks we did hook! No sooner was the grapnel down than the ship was anchored; and then came such a business: ship's engines going, deck engine thundering, belt slipping, tear of breaking ropes; actually breaking grapnels.'[44]

17 *Splicing the Cable (after the First Accident) on Board the Great Eastern, July 25th, 1865*, by Robert Charles Dudley, 1865–66.

When the cable was finally brought up from depths as great as 2,000 fathoms (3,658 m), Jenkin was surprised to discover coral formations attached to it. With an appreciation for science, he carefully recorded his findings and collected the marine animals for later examination. Charles Wyville Thomson, Professor of Natural History at the University of Edinburgh, later wrote that 'an example of *Caryophyllia*, a true coral was found naturally attached to the cable at the point where it gave way; that is to say, at the bottom in 1,200 fathoms of water' (image 18). As the cable defined the depth at which the coral was living, Thomson believed this observation was 'the final solution of the vexed question of the existence of animal life at depths in the sea'.[45] For many other scientists, a few corals attached to a telegraph cable did not topple their long-held idea of a lifeless abyss and the debate persisted.

Jenkin's corals encouraged naturalists such as Norwegian theologist and biologist Michael Sars to continue their own searches for life in the deep ocean. His son Georg Ossian Sars was also an accomplished marine scientist and together they looked to the deep regions off the coast of Norway to further disprove Forbes's theory. To reach such depths, Ossian Sars designed a new dredge; its smaller size,

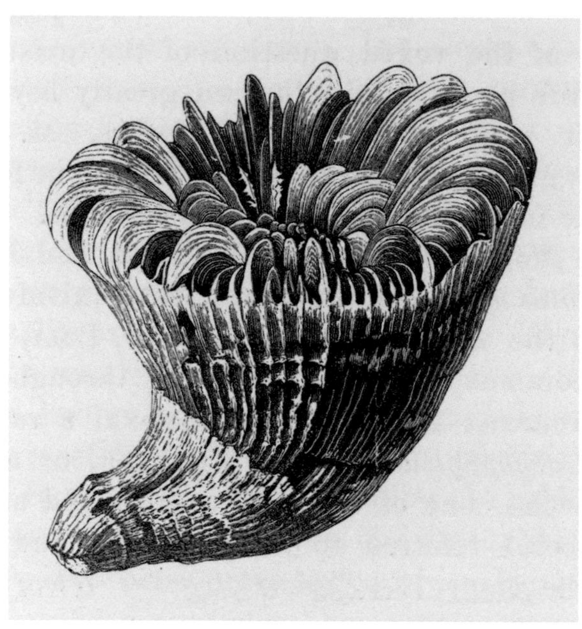

18 Illustration of *Caryophyllia borealis*, found on a submarine telegraph cable raised from 1,200 fathoms in 1860.

combined with weights, made it more suitable for collecting specimens below 200 fathoms (366 m). The device proved successful and they began building a catalogue of hundreds of species living at depths of between 200 and 300 fathoms (366–548 m) in the North Atlantic.[46]

In 1864 a breakthrough occurred when Ossian Sars — then investigating the life cycle of cod for the Norwegian government — retrieved a stalked crinoid from a depth of 300 fathoms near the remote Lofoten Islands (image 19).[47] His work contributed additional evidence to refute the azoic view. Furthermore, the ocean held life forms that scientists had not anticipated. Stalked crinoids, commonly known as sea lilies, are closely related to sea urchins and starfish; they are filter feeders that live on or near the ocean floor. At the time of Ossian Sars's discovery, naturalists easily recognised these animals from fossils. As no living sea lilies had been recorded, these types of crinoid were presumed extinct. Coinciding with discussion and debate surrounding Darwin's theory of evolution, the detection of 'living fossils' directed scientific attention to the ocean depths.[48]

Thomson shared the Norwegian naturalists' interest in crinoids and held 'a profound conviction that the land of promise for the naturalist ... was the bottom of the deep sea'.[49] As part of a small but well-connected group of European and American naturalists who studied deep-sea animals and phenomena, Thomson viewed some of Sars's 'treasures' in 1867. In Ireland, Thomson went on to study further the structure and development of crinoids and worked with his friend and colleague William Benjamin Carpenter. Inspired by the implications of Sars's collection, they discussed organising a British deep-sea expedition. Thomson asked Carpenter to use his influence at the Royal Society to convince the Admiralty.[50] While deep-sea sounding was becoming more commonplace, it would be the first

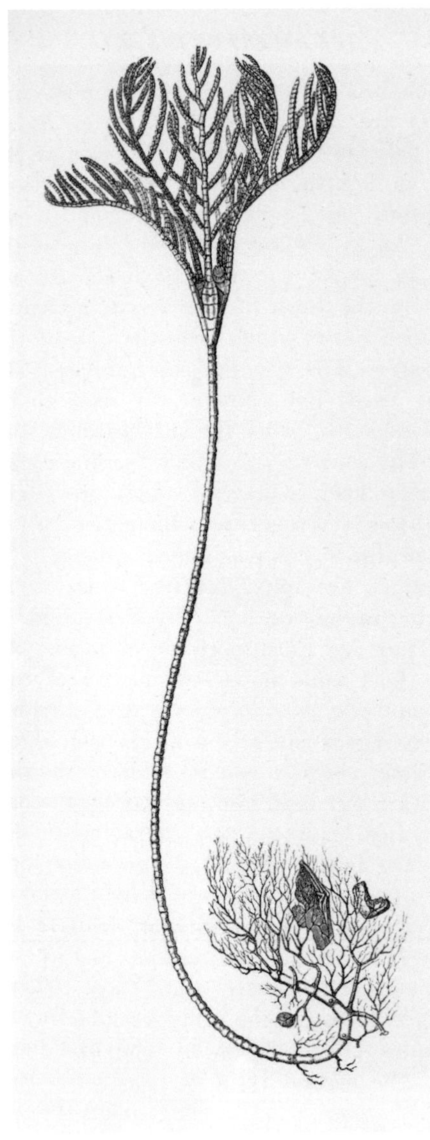

19 Stalked crinoid discovered in 1864 by Michael and George Ossian Sars.

attempt by naturalists to conduct deep-sea dredging from a British naval vessel.

By the late 1860s, diverse interests had begun to converge in the study of the deep sea. Scientists were interested in the presence and evolution of marine life, the global flow of ocean currents and temperatures, and the origin of deep-sea sediments. Meanwhile, the government desired to extend the 'red line', a network of British-controlled telegraph cables to support military operations, colonial governance and commercial shipping. This explains why, in the summer of 1868, the Hydrographic Office accommodated the Royal Society's request for Carpenter and Thomson to accompany the steamer HMS *Lightning* on a short survey cruise (image 20). Under Captain Richard Hoskyn, *Lightning* was charged with sounding the sea floor west of Ireland in preparation for the laying of additional telegraph cables.[51] Harking back to the work of Stimpson on the US Exploring Expedition of 1853–56 and aided by new techniques developed by Sars and others, the two naturalists oversaw the operation of deep-sea dredging for specimens and materials from the ocean floor.

The working conditions on *Lightning* were often dismal. Leaving Pembroke, Wales, on 4 August 1868, the naturalists set off on what would be a novel experience. For six weeks, the ageing paddle steamer battled heavy swells and wind in the North Atlantic near the Faroe Islands, a region some 140 miles (230 km) north-north-west of mainland Scotland then frequented by fishing vessels in search of cod (image 21). The naval crew deployed dredging tackle and a 'donkey engine', a small steam engine on deck, to bring up animals from the seabed. Among the hauls of sponges, echinoderms, crustaceans and molluscs, Thomson delighted in finding

20 HMS *Lightning*, a paddle steamer launched in 1823. Model by Captain John Roe, c.1979

'a magnificent specimen of a new starfish' with long and delicate arms of a 'rich crimson colour, passing into orange-scarlet' that Sars later named *Brisinga coronate* (image 22).[52]

The work was not always productive or easy. Dredging on the rocky ground of the Faroe Banks, Thomson described that the strain on the dredge ropes became so great that 'we lost two of our best dredges and some hundreds of fathoms of rope'. Using sounders and temperature gauges, they charted a bottom mixture of sand and gravel down to a depth of 650 fathoms (1,189 m) and noticed a 'cold area' of water with a temperature slightly above 0°C (32°F) found below 300 fathoms (548 m).[53] Overall, the dredging results were more than they could have hoped for. Afterwards, Thomson declared that the *Lightning*'s work showed 'beyond question that animal life is varied and abundant, represented by all the invertebrate groups, at depths in the ocean down to 650 fathoms at least, notwithstanding the extraordinary conditions to which animals are there exposed'.[54] It was a sign of the times that while surveying for the laying of telegraph cables, the *Lightning* expedition presented robust evidence against the azoic theory.

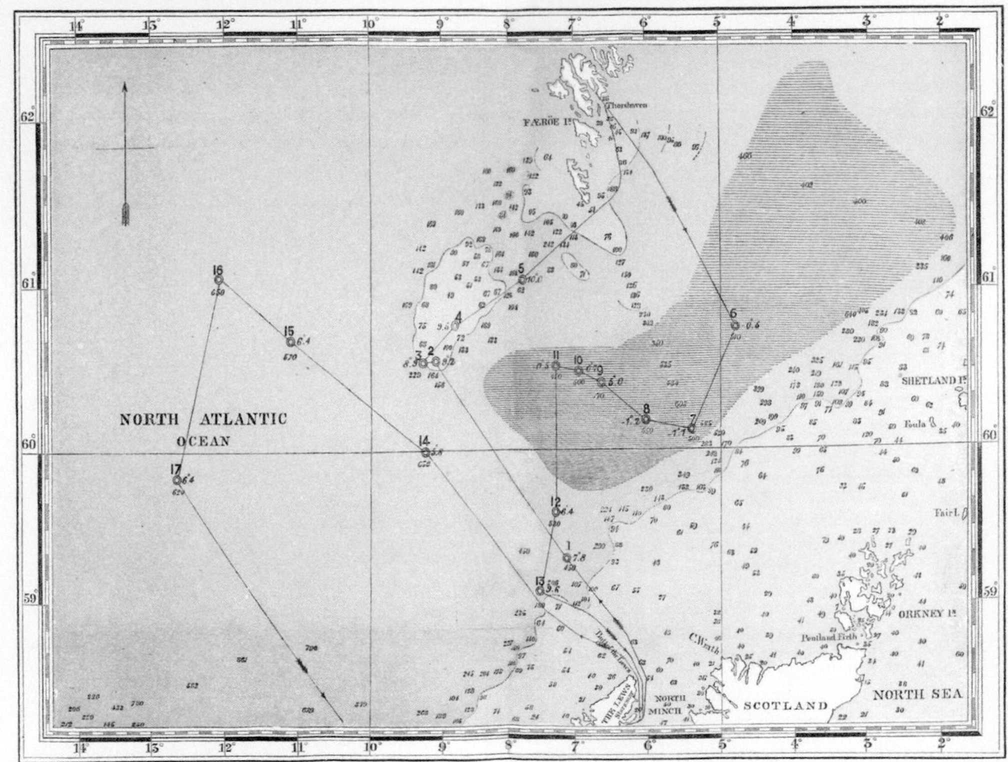

21 The route of HMS *Lightning* in the North Atlantic, 1868.

In consideration of these valuable results, the Royal Society again requested from the Admiralty the use of a steamship and naval crew. The following year, the steam survey ship HMS *Porcupine* was deployed under Captain Edward Killwick Calver to this unique service. The *Porcupine* cruises in the summers of 1869 and 1870 had wider reach, conducting a series of soundings and dredgings in the North Sea but also in the Mediterranean and the Strait of Gibraltar. Thomson and Carpenter once more led the work and were joined by Gwyn Jeffreys, whose cooperation was especially valuable thanks to his thorough knowledge of the species and distribution of recent and fossil Mollusca.[55] He had experience dredging on private yachts in shallower waters; working at greater depths on a naval ship with steam power was an exciting opportunity.

The results amplified *Lightning*'s previous discoveries. On 22 July 1869, *Porcupine* being in the Bay of Biscay, Calver performed the unprecedented feat of dredging in 2,435 fathoms (4,453 m), a depth nearly equal to the height of Mont Blanc, the highest mountain in the Alps. After seven and a half hours, 168 pounds (76 kg) of samples

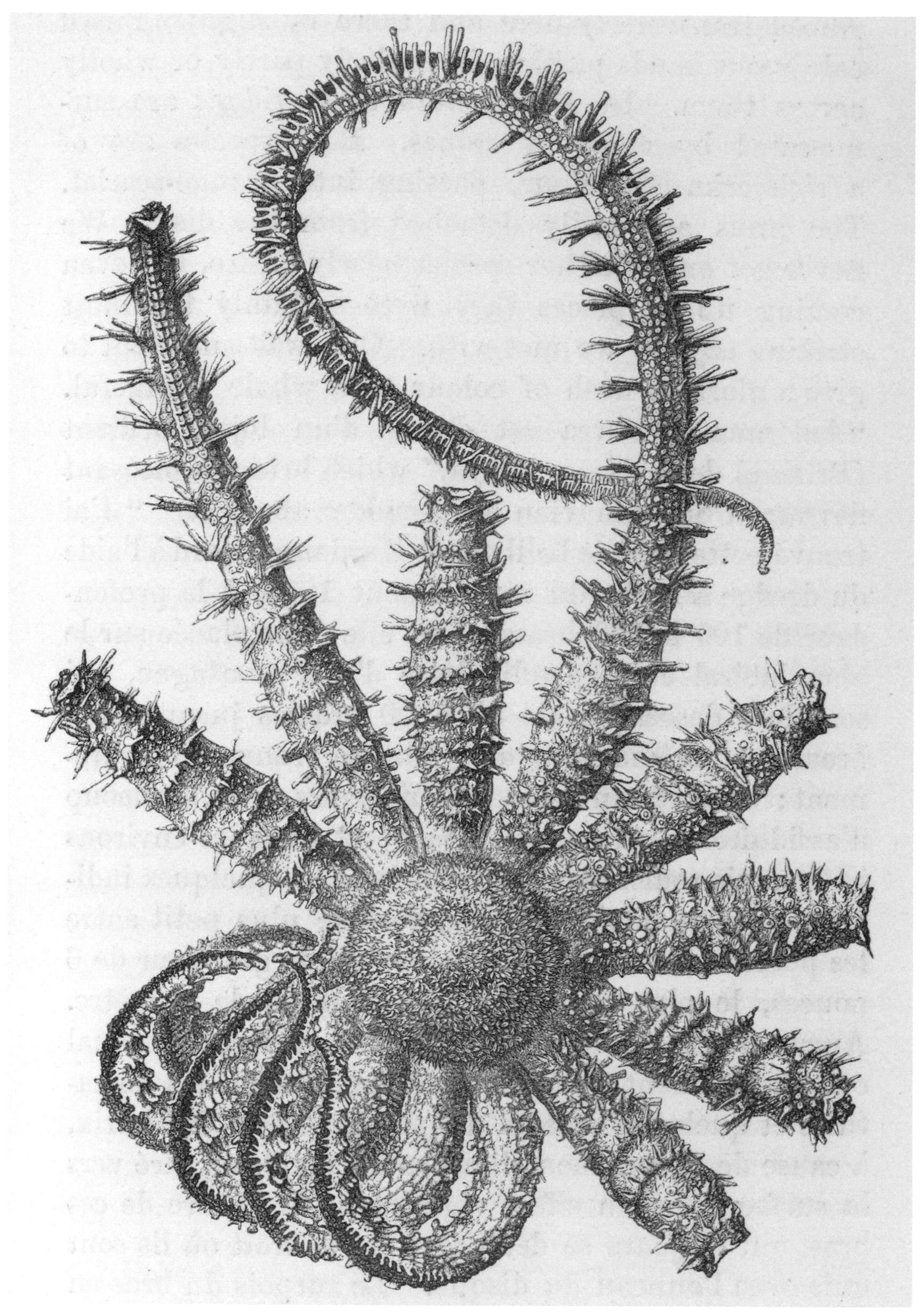

22 Illustration of *Brisinga coronate*, dredged on 3 September 1868 from a depth of 500 fathoms.

23 Dredge with 'hempen tangles' used by the crew of HMS *Porcupine*.

were raised to the surface. It was frequently found that, 'while few objects of interest were brought up within the dredge, many living creatures came up sticking to the outside of the dredge-bag, and even to the first few fathoms of the dredge-rope'.[56] To remedy this problem Calver devised an arrangement not unlike that employed by the coral fishers of the Mediterranean, who gathered a type of precious red coral that grows fixed to rocks at a depth of 60 to 80 fathoms (110–146 m).[57] He fastened to the dredge bag half a dozen swabs, made of long, coarse, hempen yarns, such as those used for drying decks (image 23). The swabs dragged across the surface of the mud and entangled the creatures living there: 'multitudes of which, twisted up in the strands of the swabs, were brought to the surface with the dredge'.[58]

After the voyage, the naturalists described many new species, including *Porocidaris purpurata*, a sea urchin that delighted Thomson with its beautiful deep purple and rose-pink spines (image 24). With proof of life now found below 2,000 fathoms (3,658 m), the *Lightning* and *Porcupine* collections generated enthusiasm for further deep-sea research both in Britain and abroad.

24 A modern specimen of *Porocidaris purpurata*, showcasing the urchin's vibrant pink colouring.

CHAPTER 2

From Warship to Research Vessel: HMS *Challenger*

> The progress of oceanography depends on a great extent upon the development of mechanical aids, by which we mean not only the scientific instruments employed, but also the whole arrangements of the ship itself.
>
> John Murray[1]

After the *Lightning* and *Porcupine* voyages in the summers of 1868, 1869 and 1870 in the North Atlantic and the Mediterranean, Charles Wyville Thomson and William Benjamin Carpenter were optimistic about continuing their deep-sea research. With the aid of disciplined naval crews, trained hydrographic officers and steam power, the ocean floor had been dredged and animal life recovered well beyond the continental shelf, at depths not attempted before. Routine deep-sea experiments, Thomson argued, could now be undertaken in earnest. In 1871 the two scientists conceived 'of the idea of a great exploring expedition that should circumnavigate the globe' with the aim to 'find out the most profound abysses of the ocean, and extract from them some sign of what went on at the greatest depths'.[2] Before a circumnavigation voyage could be a possibility, however, they first had to convince the Admiralty to dedicate a ship and crew to the task that could withstand the rigours of working for months at sea, and the government and Royal Society in London to provide everything needed for the expedition.

The Next Step: The Circumnavigation Expedition

A circumnavigation voyage would be extremely costly and the use of government resources had to be justified. To gain political support for the endeavour, Carpenter once again reached out to his influential friends at the Admiralty and the Royal Society. In a shrewd manoeuvre, he appealed to British national pride. In June 1871 he wrote to George Gabriel Stokes, Secretary of the Royal Society, emphasising that other nations were now 'entering upon the Physical and Biological Exploration of the Deep Sea' and that 'the time is now come' for Britain to initiate 'a more complete and systematic course of research than we have yet had the means of prosecuting'.[3]

Without an immediate response, Carpenter argued, other nations would claim the scientific prestige and strategic knowledge of the deep. Along with the deep-sea explorations underway by Germany and Norway, in 1871 the United States was preparing to launch the *Hassler* Expedition to conduct experiments and dredge off the coast of South America. Carpenter reminded the Royal Society that a British voyage in this realm would be welcomed 'in the scientific world and (I have good reason to believe) by the public generally' and stressed, 'No time ought now to be lost if the matter is to be taken up at all.'[4]

To get the funds approved, Carpenter acted as a middleman between the Royal Society and the Admiralty. After encouraging correspondence with First Lord of the Admiralty George Goschen, he felt that in September 1871 there was growing support for deep-sea research and the Royal Society should make its case for an exploring expedition. With this encouragement, the Royal Society formed a committee to apply for funds formally, which brought together the three original researchers from the *Lightning* and *Porcupine* cruises — Thomson, Carpenter and Gwyn Jeffreys — with notable scientists including botanist Joseph Hooker, naturalist Thomas Huxley, electrical engineer William Siemens and Hydrographer of the Admiralty George Henry Richards.[5]

The committee recommended a wide-ranging programme of scientific research. Besides scientists, naval hydrographers had an active role in hammering out the expedition's aims. Over two months, Thomson recalled that the group, 'with Admiral Richards as one of its most influential members, met from time to time and offered practical suggestions' on the voyage's preparation.[6] The following priorities were outlined: the study of the physical and chemical characteristics of the deep sea; the nature of deep-sea deposits; and the distribution of organic life throughout the great ocean basins.[7]

In the final application to the government, the President and the Council of the Royal Society asked for three provisions to carry out research effectively:

1 A ship of sufficient size to afford accommodation and storage-room for sea-voyages of considerable length and probable absence of four years.
2 A staff of scientific men qualified to take charge of the several branches of investigation.
3 A supply of everything necessary for the collection of the objects of research, for the prosecution of the physical and chemical investigations, and for the study and preservation of the specimens of organic life.[8]

These were not modest requests and there was no guarantee that the government would agree to pay such a large sum, but, with Admiral Richards's assurances and involvement, the Admiralty approved the committee's application.[9] With funding finally secured, Thomson and Carpenter's plan for an impressive circumnavigation voyage now began to take material form.

The Transformation of HMS *Challenger*

Challenger's story before the scientific expedition is not often told, but the ship's history offers a valuable window onto how the Royal Navy operated in the second half of the nineteenth century and how this shaped the oceanographic voyage. Located along the River Thames, Woolwich Dockyard was a centre of innovative marine steam engineering, and it was from there that HMS *Challenger* was first launched on 13 February 1858 (image 25). Only two weeks earlier, on 31 January, the *Great Eastern*, designed by Isambard Kingdom Brunel, had been launched, heralding a new maritime age that belonged to steam and iron. Steam engines offered strategic advantages: steamships could leave a harbour against the tide and travel independently of the wind. In battle, these vessels gained

25 HMS *Challenger* under sail during the oceanographic expedition's visit to Antarctica, attributed to William Frederick Mitchell, 1880.

superior flexibility of course and speed. Nevertheless, steam power had its drawbacks: coal was an extremely dirty, bulky, solid fuel and supplies had to be replenished regularly.

In the 1850s, motivated by escalating tensions with France, Britain rapidly converted large numbers of sailing vessels to steam power. *Challenger* was one of ten warships of the *Pearl* class of 21-gun corvettes and was built to carry a normal complement of 290 men.[10] Named after the lead ship of the class, launched in 1855, they were a hybrid of sail and steam. The application of steam propulsion significantly raised shipbuilding costs and the installation of *Challenger*'s engine and machinery amounted to £24,264, a third of the total cost of construction.

During its 14-year service as a warship, *Challenger* protected the trade and colonial interests of the British Empire. As a result of the resources put into shipbuilding, by the early 1870s over 100 Royal Navy warships were assigned to overseas stations.[11] *Challenger*'s role in defending British interests in America and Oceania were considered unremarkable for the time but are a stark reminder of nineteenth-century imperial actions. The ship's first commission was under Captain John James Kennedy, who was assigned to the North

American and West Indies station. In 1862 *Challenger* took part in operations against Mexico, including a British, French and Spanish occupation of Veracruz to enforce the country's debt repayment to Europe. During its second assignment as the flagship of the Australia station, *Challenger* was deployed by the Royal Navy to enforce colonial authority over Indigenous Peoples, sometimes with tragic consequences. One specific incident occurred in 1868. At the request of the acting British consul to Fiji, John Bates Thurston, *Challenger*, commanded by Commodore Rowley Lambert, undertook a punitive mission against the Wainimala, a clan living in the mountains on the island of Viti Levu. In the 1860s, the island had experienced an influx of white settlers from Australia and the United States, who obtained land for cotton farming, often by forceful or dishonest means. In apparent retribution for the murder of a Christian missionary, an armed force of 87 men from *Challenger* shelled and burnt the village of Deoka, resulting in the deaths of more than 40 Wainimala.[12]

The navy's large fleet meant that vessels could be recalled and redeployed where needed, and *Challenger* returned to Sheerness in March 1871. HMS *Clio*, another *Pearl*-class corvette, became the new flagship of the Australia station and *Challenger* was selected for the oceanographic expedition. *Challenger*'s physical attributes appealed to the expedition's needs. At a length of 226 feet (68.9 m) overall, 200 feet (61 m) on deck and with a beam of 40 feet 6-inches (12.3 m), the ship had ample space below deck and therefore plenty of room for scientific work. With a displacement (a calculation of a ship's weight) of 2,306 tons, *Challenger* was much larger than *Lightning*, the paddle steamer selected for the 1868 deep-sea dredging cruise in the North Atlantic, which had a displacement of 349 tons and capacity for a crew of 20.[13]

To ready *Challenger* for scientific work, the ship's refit commenced at the Royal Navy's Sheerness Dockyard, located at the mouth of the River Medway, Kent, in June 1872. The Admiralty estimated that repairs and modifications would require six months of planning and work, and it was hoped that the expedition could be ready to leave England by the end of November 1872.[14]

As *Challenger* was being prepared, the Lords of the Admiralty appointed Captain George Strong Nares, an experienced survey officer and author, to take command (image 26). After attending the Royal Naval School in New Cross, south London, he entered the Royal

26 Officers and scientists on board HMS *Challenger*, by Caleb Newbold, 1873. Sitting, left to right: Commander John Fiot Lee Pearse Maclear, Surgeon George Maclean, Navigating Sub-Lieutenant Arthur Havergal, Professor Charles Wyville Thomson, Engineer William James Joseph Spry, Captain George Strong Nares, Staff Surgeon Alexander Crosby, Lieutenant George R. Bethel. Standing, left to right: Sub-Lieutenant Andrew F. Balfour, naturalist Rudolf von Willemoes-Suhm (behind pole), chemist John Young Buchanan, Lieutenant Pelham Aldrich, Assistant Engineer William A. Howlett, artist John James Wild, Navigating Sub-Lieutenant Herbert Swire, Assistant Paymaster John Hynes, naturalist Henry Moseley, Sub-Lieutenant Arthur Channer, Paymaster R.R.A. Richards, Sub-Lieutenant Henry Cuthbert Eagles Harston, naturalist John Murray, Lieutenant Arthur C. B. Bromley.

Navy when he was 14 years old. He saw his first years of service in the Pacific and in 1852 he began his Arctic career as mate in the *Resolute* under Sir Edward Belcher, the last Admiralty expedition in search of Sir John Franklin. It was during this voyage that he learnt to navigate under sail in all wind and weather conditions. Upon the return of the expedition in 1854, Nares was promoted to lieutenant and detailed first to the Mediterranean and then to active service in the Black Sea during the Crimean War. Following the war's end in 1856, he trained cadets and wrote a bestselling book, *The Naval Cadet's Guide*, which was regarded as an essential sailing manual in its day. After conducting surveys around Australia's Great Barrier Reef, in 1867 Nares was selected for the command of the *Newport*, a vessel commissioned for surveying service in the Mediterranean. The opening of the Suez Canal in 1869 prompted demand for more detailed charts of the area and he surveyed the Gulf of Suez from 1870 to 1871.[15]

With a solid navigator and surveyor to lead *Challenger*'s naval staff, the next task was to choose the Director of the Civilian Staff, who would have a great influence on the scientific programme and outcomes of the voyage. Carpenter, who was 58 years old in 1871, had been the primary driving force behind organising the expedition, but Nares questioned if the respected university lecturer and marine researcher was a suitable match for the rigours of months at sea. Throughout August, September and October 1871, Carpenter carried out a research trip organised by Admiral Richards in the surveying vessel HMS *Shearwater*, commanded by Nares, in the Mediterranean. The short voyage provided an opportunity for Nares to work with and observe Carpenter at sea.[16] In a letter to Richards, the naval commander shared his rather unfavourable impression of the London-based philosopher:

> Would he not be more useful at home than on board to receive and work up the results. I can manage the temperature & density and current work just as well as he can. Then with young collectors in each branch of nat'l history, we shall give you good results. Dr. C is too old a man for such a long absence.[17]

With Richards's influence, the Royal Society committee chose Thomson as scientific director, who at age 41 was only a year older than Nares when *Challenger* departed. While Thomson did not have Carpenter's decades of experience as an educator or administrator, he was a better match to work with *Challenger*'s captain and officers. Thomson temporarily resigned his professorship at the University of Edinburgh and with the approval of the Royal Society selected three naturalists, a chemist and an official artist to assist him. To fill the latter position, Thomson hired Swiss natural-history illustrator John James Wild, with whom he had worked closely to produce his popular and well-illustrated book *The Depths of the Sea*, which related the experiments and deep-sea life recovered by the *Lightning* and *Porcupine* cruises.[18] Likewise, Thomson looked to Wild for help documenting the results of the *Challenger* voyage.

Upper and Main Decks: Making Space for Science at Sea

The success of the expedition depended on the merging of scientific and naval cultures at sea, and *Challenger*'s refit for the oceanographic voyage aimed to prevent potential conflicts by a careful division of workspace.[19] For instance, dredging was central to the naturalists' study, but the sorting process deposited large quantities of mud on the ship's deck (image 27). To solve this problem, a raised dredging platform was installed in front of the mainmast in the central part of the ship and built level with the hammock nettings. There, 'the contents of the dredge might be emptied, so that the naturalists, while engaged in sifting the mud and preserving specimens, might not be interrupted by the seamen working the ropes' (image 28). Equally, two large shafts were fitted from the platform to the water's edge so that 'the refuse from the dredge might be thrown overboard without dirtying the decks'.[20] In this way, the naturalists and sailors were able to work around each other (image 29).

Below the platform, immense lengths of line (rope that has been assigned a specific task on a ship) dominated the upper deck. Thomson wrote, 'Hundreds of miles of line, of strength and material suited to different purposes, are reeled and coiled in every available spot on the forepart of the main-deck and elsewhere'.[21] It was used to conduct deep-sea experiments such as sounding, dredging and measuring ocean temperatures. A small steam engine of 18 horsepower (13.4 kw) installed on the upper deck aided with the hauling of the line and instruments back to the surface.

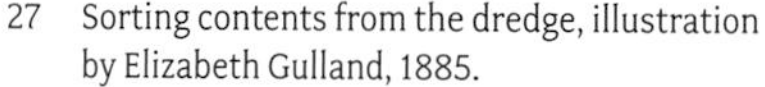

27 Sorting contents from the dredge, illustration by Elizabeth Gulland, 1885.

28 Sifting deposits from the dredge, illustration by Elizabeth Gulland, 1885.

As on the upper deck, alterations to the ship's main deck also struck a careful balance between the naval and scientific duties of the voyage (image 30). It was usual on vessels for the captain to have the largest cabin, located in the afterpart of the main deck. In *Challenger*, however, two cabins of equal size were built in this space, one each for Nares and Thomson. For the ship's company and officers, this was a bold statement that the Scottish professor held a position of authority on board and that scientific and naval objectives were both of value.

In place of its usual battery of guns on the gun deck, *Challenger* carried laboratories, workrooms and storage for marine specimens. Outside the captain's cabin Thomson described how two large workrooms were built, both '18 feet in length by 12 in breadth', one on each side of the ship: 'the room on the port side being appropriated to the use of the naturalists, whilst that on the starboard side was used by the surveying officers as a chartroom' (images 31 and 32).[22]

The natural history workroom was prepared with many of the usual instruments *Challenger* scientists would have used in a museum or university laboratory. Thomson wrote: 'The processes are much the same, only modified by the special nature of our work. We are provided with all the necessary apparatus and arrangements for skinning, mounting, and preparing specimens in all ways, and for dissecting and injecting.'[23] Special modifications were made to counter the movements of a rolling ship. For instance, racks of test tubes were fitted against the walls to prevent breakage and

29 Dredging and sounding arrangements on board *Challenger*.

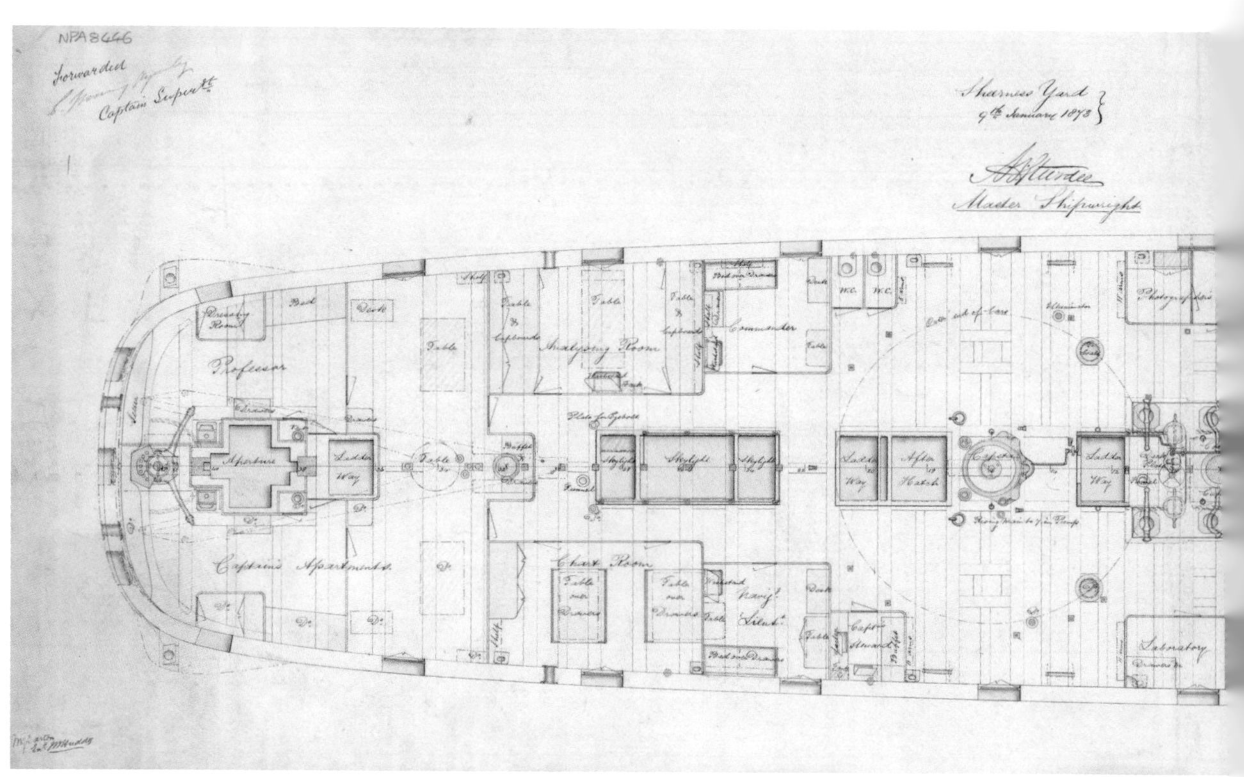

30 Main deck plan of HMS *Challenger*, as the ship was fitted for the expedition, 1873. Cabins of equal size were provided for Captain Nares and Professor Thomson.

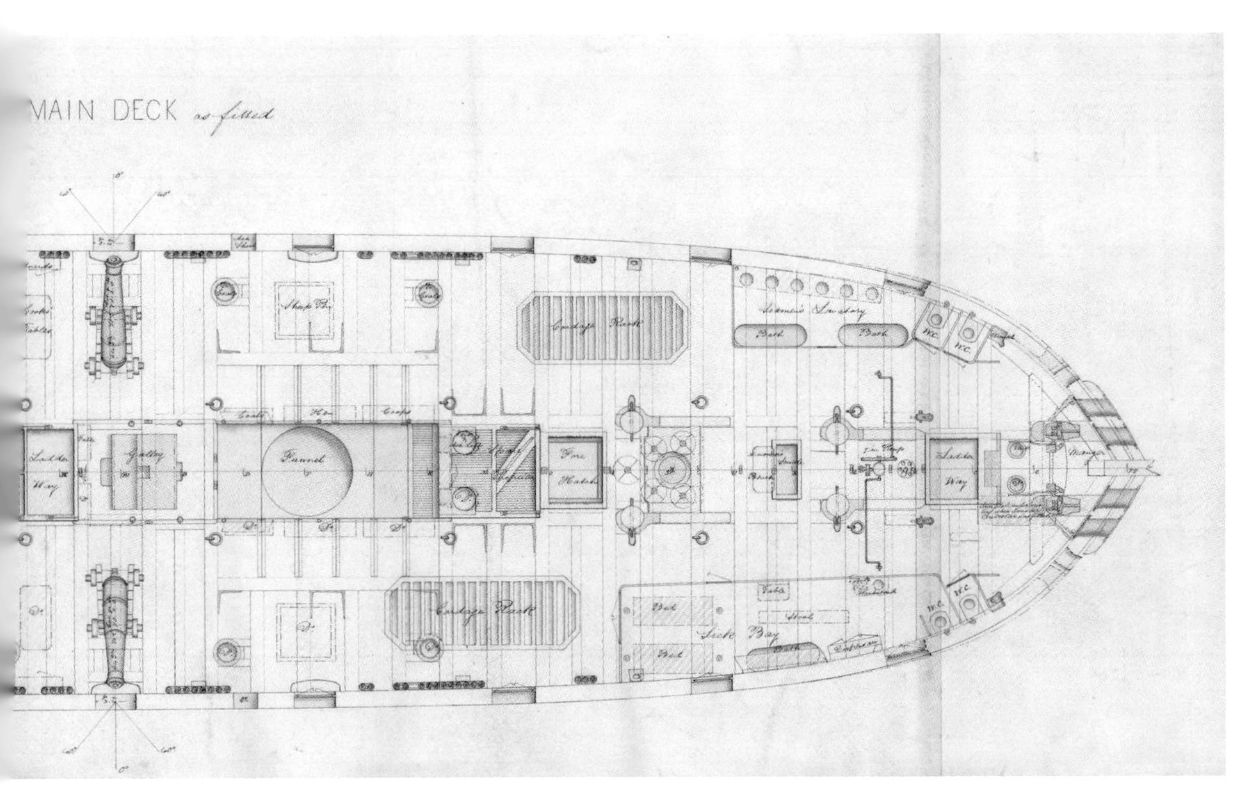
MAIN DECK as fitted

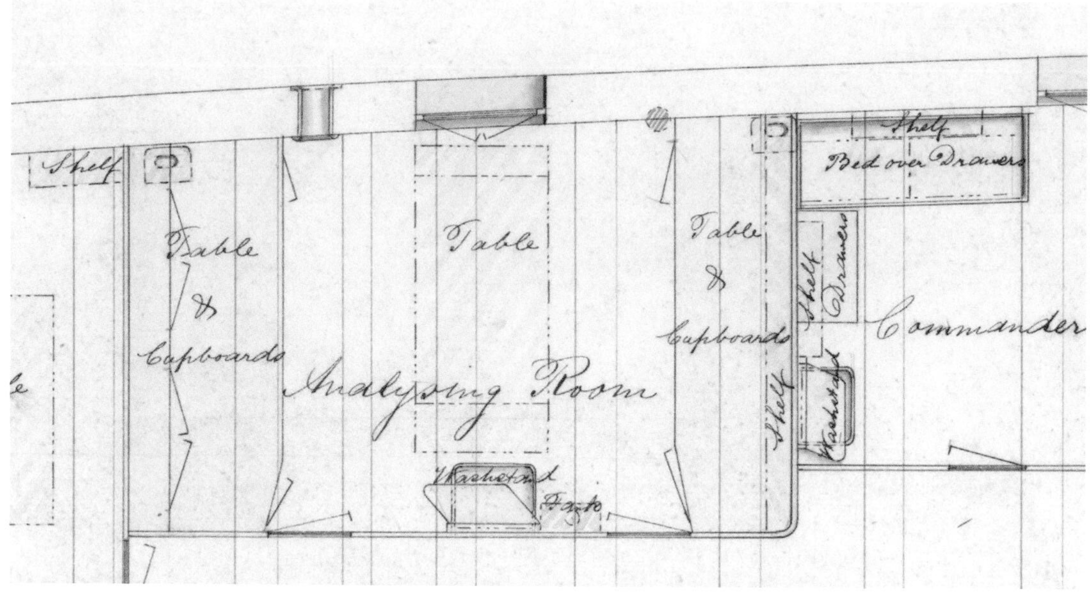

31 The analysing room on the main deck of HMS *Challenger*.

equipment, such as scissors, forceps and scalpels, was stored in small compartments in dresser drawers to stop them from knocking into each other and becoming blunt. Two microscopes were secured with clamps to a long workroom table to keep them upright at sea.[24]

The three naturalists who worked in this part of the ship hailed from different countries — Britain, Canada and Germany — and all had some experience at sea. The English botanist Henry Nottidge Moseley studied science at Oxford and medicine at University College London before leaving in 1871 to accompany the British Solar Eclipse Expedition to India. Not overly impressed by the mud or small animals *Challenger* brought up from the dredge, he was more interested in the plants and animals he discovered and collected on shore, along with writing detailed descriptions of the people and cultures the expedition encountered.

John Murray did not have Moseley's Oxford credentials, but he had voyaged in hostile environments and had a passion for ocean science. Born in Canada in 1841, Murray moved to live with his Scottish uncle as a young man. He attended the University of Edinburgh but did not complete his medical studies. Like Moseley, he left behind his formal education to explore the natural world and joined a whaling ship, the *Jan Mayan*, as its surgeon. While this was not a scientific expedition, the whaling crew spent seven months in the Arctic, during which time Murray collected marine specimens and observed the weather, ice

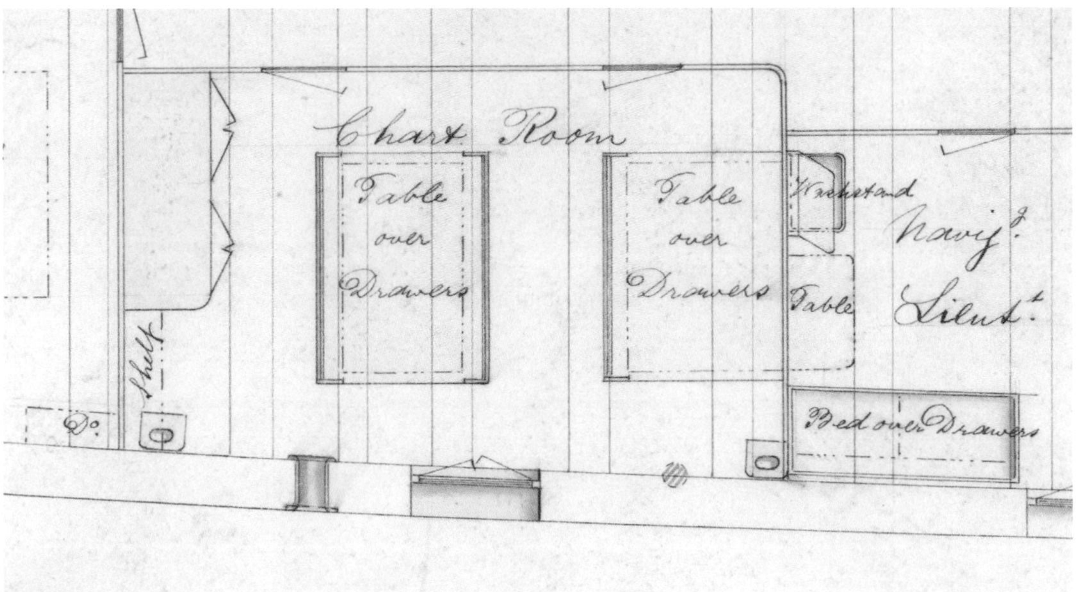

32 The chart room on the main deck of HMS *Challenger*.

and ocean currents. This experience made him a good candidate for the circumnavigation and, since Murray had studied geology during his time at university, he was selected to oversee the expedition's collection of rocks and samples from the ocean floor.

In 1871 German zoologist Rudolf von Willemoes-Suhm lectured at the University of Munich and a year later he joined his country's North Sea Expedition. During the voyage's stop at the port of Leith, near Edinburgh, he met Thomson and was quickly selected for *Challenger*.[25] The naturalist had expertise in describing and illustrating a wide range of marine life — from fish to crustaceans and worms — useful skills for an expedition setting out to determine the distribution of life throughout the great ocean basins.

Across from the natural history workroom, an equally spacious cabin on the starboard side was used by naval surveying officers as a chart room. It was installed with a complete set of charts of the world, instruments such as theodolites (used to measure precise horizontal and vertical angles), books, sailing directions, drawing materials and stationery. It was here that experienced hydrographer Navigating Lieutenant Thomas Henry Tizard, under the direction of Nares, oversaw the meticulous work of updating Admiralty charts and creating new ones with the information *Challenger* gathered, not only of shorelines but also diagrams of the topography of the ocean floor and corresponding water temperatures.

33 Illustration of the chemical laboratory on board HMS *Challenger*.

In the central part of the main deck, storage space was found for additional scientific equipment and two smaller cabins were built abreast of the main mast. Thomson described how the area between the chart room and the laboratory was 'occupied by such of our gear as would not pack into the workrooms', including a hydraulic press for testing thermometers and other instruments exposed to the great depths.[26] On the port side was a specially designed photographic laboratory and on the starboard side a chemical laboratory, both fitted with apparatus fixed and adapted for use on board (image 33).

The expedition's official chemist was John Young Buchanan, another member of the scientific team with Scottish connections. Born to a wealthy family in Glasgow, he studied chemistry at Glasgow University before continuing his education in Germany and France. Prior to joining *Challenger*, he was elected a Fellow of the Royal Society of Edinburgh, a prestigious scientific institution of which Thomson was also an active member. Although sometimes contrarian, Buchanan's work dealt with the careful analysis of ocean water samples for salinity and other properties.

Lower Deck: Social Spaces

In contrast to the extensive remodelling of the upper and main decks, Thomson stated that the 'fittings on the lower deck differed but little from the ordinary fittings of a man-of-war' (image 34).[27] Most of the ship's complement slept, ate and socialised on the lower deck. As with past expeditions, the majority of the crew was carefully selected from a large group of volunteers. Abraham Smith, a sailor from Upchurch, Kent, who had previously served in over a dozen Royal Navy vessels — including a mail boat in Shanghai and HMS *Victory* in Portsmouth — wrote in his memoirs about joining up with the deep-sea expedition:

> A notice was put up in Sheerness Barracks, saying that two hundred and fifty hands were required for H.M.S. *Challenger* and anybody wishing to volunteer their services must send in their names by a certain date. Out of six hundred who volunteered their services, two hundred and fifty were summoned to pass the medical tests and I am glad to say that I am one of those who successfully passed the four doctors.[28]

The ship plans reveal that living quarters on board *Challenger*, as on other naval vessels, carefully regulated individuals' interactions depending on their rank and position. Lower-ranking members of the crew slept in hammocks forward of the main mast, while the naval and scientific staff occupied cabins with small beds located aft.

Besides the addition of naturalists, there were other staff changes on board *Challenger*. With no intention of engaging in hostilities, the positions of ship's gunner, marine officer and chaplain remained unfilled. Furthermore, no midshipmen or subordinate officers accompanied the ship on its exploring voyage. The lack of junior officers resulted in their designated mess being demolished, making way for five cabins for the civilian staff (image 35).

Social spaces created during the refit helped to ease potential tensions between scientists and officers. All the cabins on the lower deck surrounded a wardroom, the mess where commissioned officers ate together. As part of *Challenger*'s alterations, the wardroom was significantly enlarged to accommodate the scientists. In an era when it was unusual to socialise outside one's class or position,

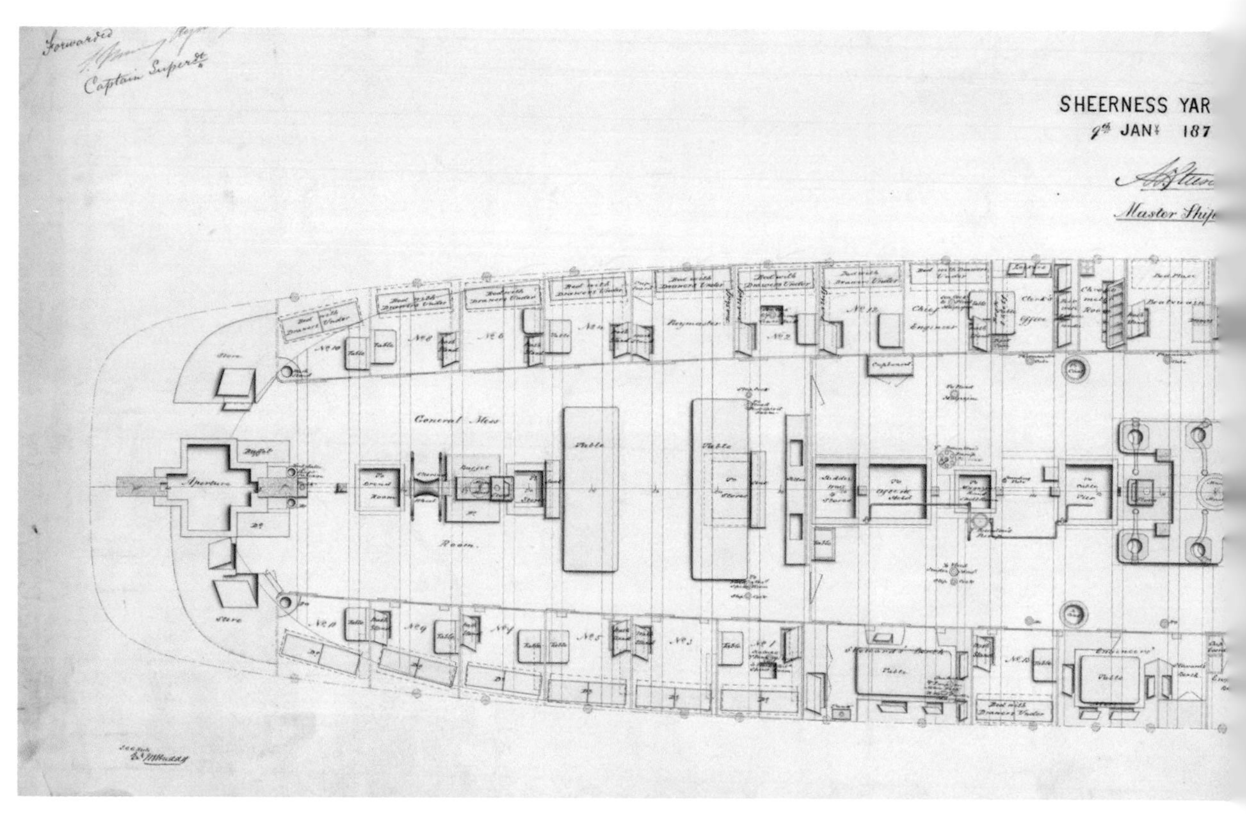

34 The lower deck of HMS *Challenger*, as the ship was fitted for the expedition, 1873.

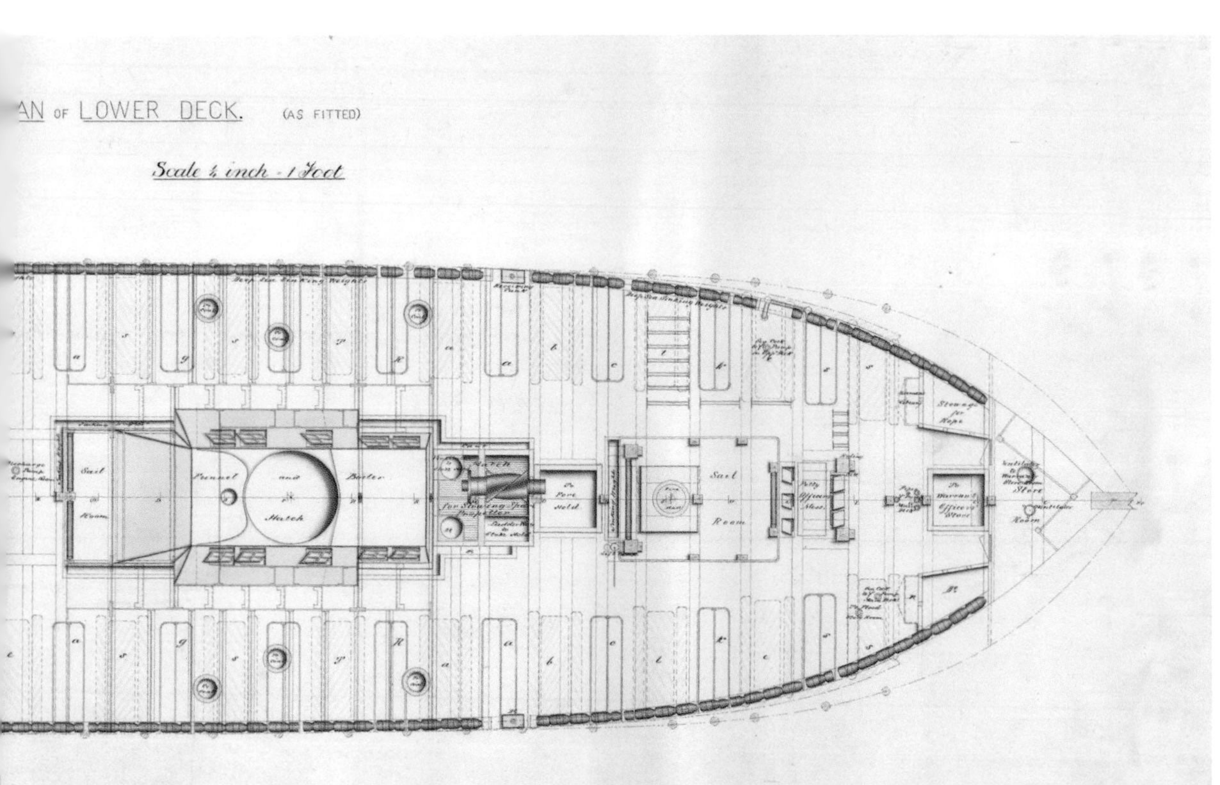
AN OF LOWER DECK. (AS FITTED)
Scale ¼ inch - 1 Foot
Sail Room
Funnel and Boiler Hatch
Fore Hold
Sail Room
Stowage for Rope

35 The naturalist's cabin on the lower deck.

the wardroom provided a space where scientists and officers could meet informally.[29] Tizard remarked that the division of space during the day 'enabled each member to work uninterruptedly at his own speciality' but that the staff met in the evenings 'to compare notes with the others in the smoking circle daily after dinner, a function always well attended, and where the events and work of the day were freely and amicably discussed'.[30]

The Hold: Steam Power

While wind power extended the ship's range, steam power was the key to unlocking the secrets of the deep. Murray declared, 'Steam first made it possible to examine properly the vast marine areas of the world.'[31] Located in the hold, the lowest deck of the ship, *Challenger*'s two-cylinder trunk engine and single-screw propeller system was an elegant piece of Victorian engineering. The furnaces raised heat, the boilers converted water to steam and the engine transformed the steam's energy into revolutions of a shaft connected to the ship's propeller. Compared to the earlier use of paddle wheels, propellers offered navigators the benefits of increased speed, stability and reliability — all of which supported deep-sea experiments such as sounding and dredging. Vessels running on steam power alone, however, were limited in their range. Steam engines also generated an enormous amount of heat that permeated throughout a ship's lower decks, a problem exacerbated when travelling in the tropics.

In operation, the steam engine was powered by dirty coal, extreme temperatures and hard human labour. *Challenger*'s four boilers were heated by a series of coal furnaces and, while steam power implies mechanical force, much of the work was done by hand. Chief Engineer James Ferguson and his team of four were responsible for overseeing the engine's output and performance. At sea, the crew, including 14 stokers and coal trimmers, followed a regular routine based on watches (bursts of activity for a set amount of time) followed by brief stretches of rest (image 36). Over a 24-hour period, *Challenger* operated a system of seven watches, beginning with the morning watch from 4 a.m. to 8 a.m. Each subsequent shift lasted four hours, except for two 'dog watches' (between 4 p.m. and 8 p.m.), which were two hours each.[32] Responding to officers' orders shouted above the din, trimmers moved tons of coal from the ship's bunkers to the boiler room. There, stokers controlled the heat of the furnaces — and thus the speed of the engine — by adding coal, regulating the air flow, raking the fires and removing ash and debris from the bottom of the furnace grates. During the engine's operation, the hold quickly became smoky, dark and scalding hot. Contemporary accounts relate that the temperature of a stoker's surroundings could vary from 120°F to 160°F (49°C to 71°C).[33]

A striking example of *Challenger*'s steam engine at work — and the amount of coal it required — can be found during the first days of the cruise. When the ship set out from Sheerness on 7 December 1872, gales in the English Channel gained strength. As the storm intensified, Nares ordered the sails to be brought in and the ship travelled under steam. Joseph Matkin, the Assistant Steward, wrote of a terrific wave that struck and damaged the ship: 'the Lifeboat cutter was smashed to atoms, and the Jib boom carried away with all the Head Sails.'[34] He described the slow progress to Portsmouth: 'the distance is 105 miles & we were 109 hours steaming full speed all the way.'[35] With the engine continuously running through the storm, the men shifted 87 tons of coal — almost a third of the ship's total

36 *Challenger*'s head stoker, Charlie Collins, can be seen standing at the back of this group (third from the left) of *Challenger* crew members with two Germans rescued from Inaccessible Island, Tristan da Cunha, by Frederick Hodgeson, *c.*October 1873.

capacity of 250 tons — into *Challenger*'s furnaces over the course of five days.[36]

Whenever feasible, the ship travelled under sail power to conserve limited coal supplies, but steaming was always necessary when carrying out oceanographic experiments. It needed to be kept as stationary in the water as possible to reduce drift and accurately measure the ocean's depth during sounding. While dredging, the vessel's propeller created the forward momentum necessary to tow a weighted net along the ocean floor. In total, *Challenger* consumed nearly 5,000 tons of coal during the voyage, a staggering amount to procure, carry into the boiler room and feed into the furnaces.

This meant that *Challenger* had to refuel regularly. In the 1870s, 'coaling' or transporting bags of coal from dock to steamship was primarily done by hand. On board *Challenger*, this work usually involved the entire ship's company and required one or two days to complete.[37] Each station had its own protocol and methods of coaling, however. At St Vincent, in the Caribbean, Tizard described that 'the coal, kept in bags, is conveyed to the ships in barges, and labourers can be hired from the shore to assist in passing the bags on board' so that vessels could continue their voyages without delay.[38]

Essential as it was to *Challenger*'s scientific operations, coal was also connected to colonial systems of power exerted by Britain and other imperial nations in the late nineteenth century. The Royal Navy generally used the whole crew to coal, but the strenuous and dirty work of coaling, essential to keep the world's growing number of steamships moving, was often done by the poorest in society, including forced labourers. While sailing around the Philippine Islands, *Challenger* received 100 tons of coal at Port Isabel, a Spanish colony on the Zamboanga Peninsula. Matkin related that 'the convicts carried our coal on board; they were rigidly overlooked by Spanish soldiers'.[39] In another instance, at the port of Bahia, Brazil, he wrote that enslaved people carried coal onto *Challenger*: 'We had about 60 of them getting in our coal while the ship's Company went on leave, and they got in 200 tons in a day.'[40] While steam power was a great advantage in the early exploration of the ocean depths, coal had a human cost.

By the 1870s, a global network of naval bases and coaling stations transported and dispensed coal among distant points of the British Empire to power the Royal Navy fleet. As ocean science depended on a steady supply of coal, those same coaling stations significantly shaped *Challenger*'s study of the sea and the areas of the ocean that were investigated.

The British naval base at Gibraltar was one of the established coaling stations upon which Nares relied during the circumnavigation. Located at the Iberian Peninsula's southern tip, the fortified base at the gateway to the Mediterranean fuelled ships on the way to southern Europe, Africa and the Suez Canal. Engineer William Spry noted the importance of the station to the fleet: 'Stores of all descriptions are to be obtained, and large quantities of coal, some 10,000 or 15,000 tons, are usually on hand.'[41] During *Challenger*'s visit in January 1873, Matkin wrote, 'We came alongside the jetty on the South Mole as soon as the Fleet had gone ready for coaling, and today, Monday, all hands are busy filling up with it.'[42]

Naval bases alone could not satisfy the British fleet's need for coal. Like other steamships in the 1870s, *Challenger* routinely visited coaling stations owned by private companies or other navies. For instance, *Challenger* arrived at Lisbon on 3 January 1873 and took on 150 tons of coal to continue dredging operations.[43] Foreign coaling stations such as that in Lisbon significantly extended the range of

37 *Challenger* hosted a visit by the King of Portugal, Luís I, an important diplomatic event, by Caleb Newbold, *c.*3–12 January 1873.

British steamships, making coal a vital aspect of diplomatic relations. During *Challenger*'s visit, the officers and scientists greeted the King of Portugal, Luís I, his officers and the British ambassador (image 37). Matkin described the meeting as a 'grand affair' with lunch. 'All of our officers were in full uniform & the ships company were all in their best, & manned the yards for him.'[44]

Challenger's Route: Maintaining the Ship and Crew

The same resources that supported the British Empire in projecting naval power in different regions of the world enabled *Challenger* to conduct its global oceanographic study. During the 1872–76 voyage, the ship travelled throughout the North and South Atlantic; crossed the Antarctic Circle and encountered icebergs; and then sailed onwards to Australia. Exploring the ocean along the western regions of the Pacific, *Challenger* visited Fiji, New Guinea and the Philippines before stopping at Hong Kong to make repairs and spending three months in Japan. The next leg of the journey involved investigating the middle of the Pacific, navigating from north of Hawai'i to south of Tahiti. Finally, *Challenger* passed through the Strait of Magellan before travelling across the Atlantic once more on its northbound voyage to England (see image 132, pp. 228–29).

Throughout the three-and-a-half-year voyage, places on land regularly provided *Challenger* with everything that the expedition needed: from hull and engine repairs to vast quantities of fresh and preserved food that sustained the crew for several weeks at sea. As with coal reserves, *Challenger* trusted Royal Navy bases to deliver the bulk of its supplies. After being battered in the English Channel and repaired at Portsmouth, the ship's route coincided with visits to Royal Navy bases — extensive naval complexes that supported Britain's fleet — located at Gibraltar, Halifax, Bermuda, the Cape of Good Hope, Sydney and Hong Kong (image 38). *Challenger* was studying the ocean, but it followed a supply route that connected British military bases around the globe.

38 HMS *Challenger* at the naval base at Bermuda, West Indies, by Caleb Newbold, *c.*4 April–31 May 1873.

High-performance steamships demanded extensive maintenance, without which the expedition could not operate. Engines required skilled repairs and precision parts, and the hulls of large vessels could only be cleaned and repaired in dry dock. Typical of a visit to a naval station, when *Challenger* arrived at Sydney, the expedition obtained coal, conducted repairs and took on large quantities of supplies. Matkin wrote: 'at present [we] are busy taking in six months provisions to last us to Hong Kong next November'.[45] In addition, the ship was sent to Fitzroy Dock on Cockatoo Island, one of the few dry docks in the South Pacific, to check the condition of its hull.

Of course, the ship relied on more than technology and supplies to continue its voyage around the globe. The ship's company managed the heavy rigging and sails, conducting all the work necessary to maintain the ship at sea. *Challenger* was not an easy assignment, even for experienced naval hands. Normally, naval vessels travelled quickly from one port to another, but conducting deep-sea sounding, dredging and temperature readings lengthened significantly the time spent at sea. Smith remembered 'the longest time that we were at sea without sighting any land, with only the sea and sky to look at was three months',[46] and, in total, the expedition travelled a distance of 68,890 nautical miles (127,580 km) with 719 days at sea.[47]

As on modern oceanographic expeditions, *Challenger*'s crew found ways to entertain themselves and keep up spirits, including dancing, music and putting on plays. Matkin described *Challenger*'s brass band composed of seamen and marines (image 39), who were enthusiastic, if not experienced, musicians:

> The Officers bought the Instruments & provided a Bandmaster to teach them; they were 15 volunteers & 9 wanted to play the big drum, they practise every day in the fore peak of the vessel & the noise is something fearful & causes the Watch below to swear a good deal. The Bandmaster expects to fetch tolerable music in about 6 months.[48]

In early March 1873, as the ship was completing its first Atlantic transit and approaching the Caribbean island of St Thomas, Matkin reported that the band 'have gotten on very well indeed' with their practice and for the first time the ship's band performed selections for the crew, including two marches and 'God Save the Queen'.

39 The eleven regular seamen who made up the *Challenger* band pose on deck with drums, brass instruments, and a cat. Photographer and date unknown.

When they had finished, 'the Ships company clapped them as if it was the first time of hearing — instead of being treated with the same tune every day for the last 3 months.'[49] By the time the ship had been at sea a year, the band could play very well; they performed when the ship left harbours and on festive occasions, and even for visiting dignitaries.

Music and other diversions may have made the voyage a bit more pleasant but, of the ship's original company of 243, only 144 men remained on board for the entire circumnavigation, a testament to the harsh conditions.[50] Together with food, coal and supplies, the expedition had to replace several crew members along the route. Seven deaths occurred during the cruise: two were due to natural causes, three were caused by violence and two by what the ship's surgeon George Maclean described as 'acute poisoning'. Naturalist Willemoes-Suhm perished in the Pacific from erysipelas, an infection of the upper layers of the skin, and *Challenger*'s first schoolmaster died 'of a disease of the brain'. Two men drowned during the voyage and a young boy suffered a fatal blow to the head while dredging.

One man perished from alcoholism and another committed suicide by drinking chloral hydrate, a drug administered as a sedative, sometimes before surgery.[51]

In addition to the fatalities, another 29 men left the ship due to illness, discharge, imprisonment or reassignment.[52] The most notable disembarkation occurred when Captain Nares departed at Hong Kong to lead the British Arctic Expedition of 1875–76. Matkin wrote that the crew was 'all sorry to lose him for he was a very kind & good man' and Nares was sent off with all hands mustered in the rigging, the band playing and '3 good cheers'.[53] On 2 January 1875, Nares was replaced by Captain Frank Tourle Thomson, who had recently taken command of a wooden screw corvette, HMS *Modeste*, at the China station. Thomson, who had once served as Captain of the Royal Naval College, Greenwich, and on the royal yacht *Victoria and Albert*, lacked Nares's knowledge of deep-sea dredging and sounding and had not participated in scientific voyages. Matkin remarked on the new captain's fondness of cigars, playing piano and 'good living' but had few complaints.[54] Aided by Tizard and the remaining staff, he continued *Challenger*'s regular programme of oceanographic experiments, established during the first half of the circumnavigation.

The greatest reduction in the ship's company, however, was caused by the desertion of 60 men at ports of call.[55] *Challenger* conducted sounding and other oceanographic experiments at 504 stations during the voyage, all adding to the crew's workload. Matkin thought that 'the work is much harder for everyone than it is in an ordinary man of war whilst the pay is the same' and noted a dispiriting event in which two band members deserted at Sydney.[56] For those who did stay on board, *Challenger* made a grand return to Britain three and a half years after its departure. Smith recalled:

> When we sighted the Channel Fleet, they were in two lines, and the Admiral signalled to our Captain to steam down the middle which we did. All the officers and crew of the Fleet cheered us as we passed and the bands on all the vessels were playing. As we entered Portsmouth all the ships laying off Spithead cheered us.[57]

After stops at Portsmouth and Sheerness, the ship's company was paid off (the practice of ending an officer's commission and paying

40 HMS *Challenger* as a hulk without its masts and funnel, decommissioned by the Royal Navy and moored at Chatham Roads, by George Cochrane Kerr, *c.*1890.

the crew's wages once a ship concluded its voyage) at Chatham on 12 June 1876 and the historic voyage came to an end. *Challenger* was emptied of all its scientific instruments and stores of natural history specimens and returned to the fleet, operating as a Royal Naval Reserve training ship until 1880, when it was converted into a storage hulk in the River Medway and stayed until it was sold and broken up in 1921 (image 40).

Making space for science at sea — with special laboratories and workrooms, as well as recognising scientists as important members of the crew — was one of *Challenger*'s lasting contributions to oceanography. *Challenger*'s figurehead still exists and is now kept at the entrance to the National Oceanography Centre at Southampton, a permanent reminder of the converted warship that broke new ground in the exploration of the deep sea. Other research vessels soon followed, but, with the advance of steamships and deep-sea sounding techniques at the beginning of the twentieth century, *Challenger* was the last wooden sailing ship to conduct an oceanographic expedition on such a scale. The global infrastructure and resources of the Royal Navy made the circumnavigation possible; later expeditions would focus on smaller areas, slowly building up a more detailed picture of Earth's oceans.

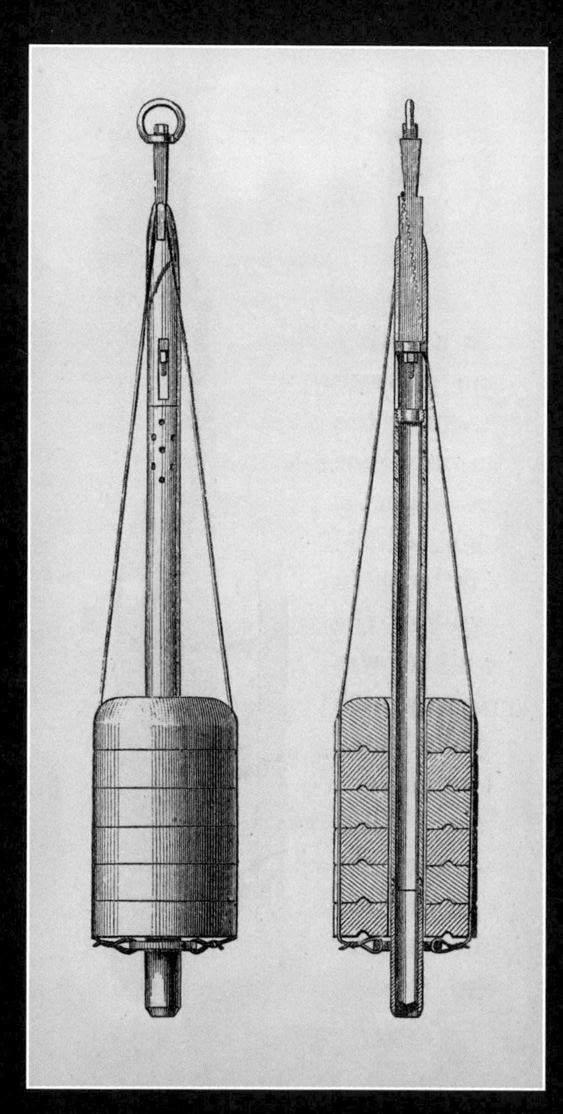

CHAPTER 3

Exploring the Depths: The Trials of the Baillie

> Studying the depths of the sea is like hovering in a balloon high above an unknown land which is hidden by clouds.
>
> John Murray[1]

Challenger's commission was immensely ambitious. Before the ship departed from Sheerness, the Hydrographic Office issued Captain George Strong Nares detailed instructions that included the expedition's route and an extensive list of experiments and observations to be carried out during the voyage. They ranged from the formidable task of determining 'the distribution of organic life' throughout the oceans *Challenger* traversed to performing hydrographical examinations 'of all the unknown or partially explored regions which you may visit'.[2] Within this imposing programme, in his role as Hydrographer of the Admiralty, George Henry Richards made the expedition's primary mission clear: 'If any one of the various objects of the expedition is more important than the other, it may be said to be the accurate determination of the depth of the ocean.'[3] The Royal Society reinforced that deep-sea sounding 'is an object of such primary importance that it should be carried out whenever possible' throughout the voyage (image 41).[4] Yet, sounding the ocean depths posed engineering problems as well as bodily risk. 'In all deep-sea investigations it is of course of the first importance to have a means of determining the depth to the last degree of accuracy,' Charles Wyville Thomson reflected, 'and this is not so easy a matter as might be at first supposed.'[5]

41 An illustration of *Challenger* shortening sail in preparation for sounding.

'If some instrument could be devised'

Sounding or determining the depth of the water beneath a vessel's hull with a 'lead and line' is one of maritime navigation's oldest practices (image 42). In shallow waters or near shore, sailors used a simple sounding method: a weight, often made of lead (since it is a heavy and durable metal, and whence it took its name), was affixed to a long rope and thrown overboard; mariners then counted how many fathoms of line, designated by coloured pieces of cloth, passed through their hands before the weight hit the bottom. To discover the composition of the seabed, tallow (a sticky substance made from animal fat) was placed in a small depression in the lead's base. On hitting the bottom, mud or other debris from the seabed stuck to the tallow. When retrieved, the 'nature' of the bottom was identified by colour, texture and even taste. These descriptive features could then be compared to charts. Together, both the depth of the water and composition of the ocean floor aided navigation and helped to locate a vessel in relation to the shore.

42 A navigator is pictured sounding the water depth using a 'lead and line' in this illustration from a 16th-century atlas called *The Mariners Mirrour*, 1588.

The investigation of the deep ocean added many complications to the practice of sounding. One of the first problems was that a sounder had to travel significant distances from a moving ship to reach the ocean floor. In a perfect reading, the line would measure the shortest distance from the vessel at the surface to the seabed directly below. The speed of travel was an important factor: if a sounder descended too slowly, the vessel's momentum and the ocean current caused the rope to drift as the instrument fell, falsely adding length to the depth measurement. One solution was to deploy a heavy weight, such as a cannonball, to a slim line that offered little resistance in the water. While this increased the accuracy of depth measurements, a light material such as twine was not strong enough to carry the sounder to the surface. Therefore, great depth could be determined but not the nature of the seabed, a problem for hydrographers and telegraphic companies.[7]

Steam power and advances in sounder design propelled a new age of deep-sea exploration. US Navy Midshipman John Brooke solved part of the engineering dilemma of deep-sea sounding by inventing the first device to abandon its weight when it hit the bottom. The Brooke's sounder had the advantage of a speedy descent to the sea floor but then, relieved of its weight, was light enough to be brought back to the surface. Furthermore, with the aid of a steam windlass

43 Mechanical marine sounder used on board HMS *Bulldog*, 1860.

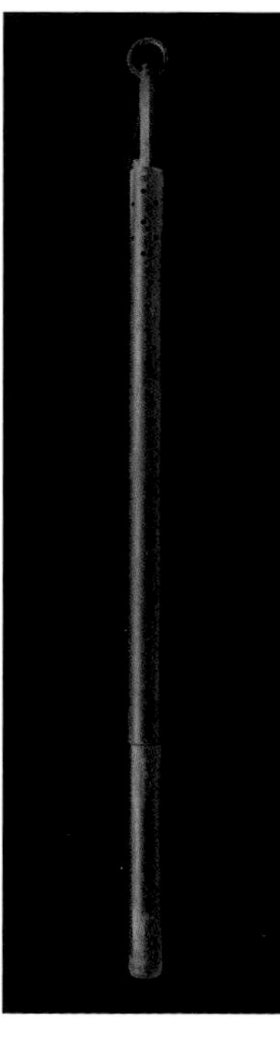

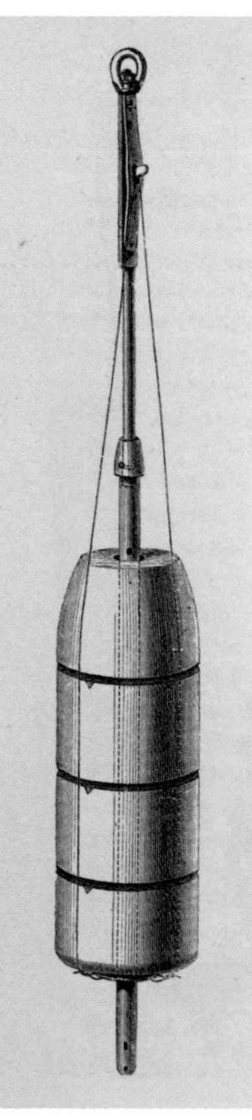

(for moving heavy weights) on deck, crews no longer had to haul sounders to the surface with human strength alone. Deep-sea sounding became more efficient and practical, and throughout the 1850s telegraph ships and naval vessels deployed sounders based on Brooke's design.

Commercial interests spurred further sounding innovations. Reviewing the failure of the first transatlantic cable, the British Submarine Telegraph Committee remarked in 1861 that telegraph companies needed more detailed information about the ocean floor. Accordingly, the committee recommended that 'it would be of great advantage for this purpose if some instrument could be devised which would enable the actual outline of the bottom of the sea to be traced'.[8] In response, throughout the 1860s and early 1870s a variety of sounders were designed and tested to survey the topography and composition of the sea floor.

To raise ooze, mud, rock and sand from the bottom of the sea required a modification of existing sounders that had been designed primarily to measure depth. One of the first sounders to collect deep-sea samples was a sounder clam, devised in 1860 by George Charles Wallich, ship's surgeon in HMS *Bulldog* (image 43). While the *Bulldog* sounder's hinged cups captured more material from the ocean floor, samples could be washed out and lost during the long ascent to the surface.

In 1868, under the command of Captain Peter Frederick Shortland, the wooden steam paddle sloop HMS *Hydra* sounded across the Arabian Gulf in preparation for laying the Indian cable.[9] During this voyage, a blacksmith and two sailors invented a sounder, which was given the ship's name subsequently. While derived from the Brooke's sounder, the Hydra's sounding rod was hollow and divided by butterfly valves (images 44 and 45).

44 (above) This Hydra sounder may have been one of the six used when the *Challenger* expedition began in 1872.

45 (below) The 'Hydra sounding machine'. The brass cylinder had a butterfly valve at the bottom of the tube and a sliding iron rod at the top; weights were supported by a wire attached to a washer.

When hitting the sea floor, the weight of the sounder drove the end of the cylinder into the ground; a spring then released the sinkers. When the sounder was hauled up, the butterfly valves closed and trapped a sample of the seabed in the cylinder, reducing the amount of material washed out during retrieval.[10] The Hydra thus had an advantage over previous designs in that it could pierce the ocean floor, capture a sample of sediment in a tube and carry it back to the surface for collection.

In December 1872, when *Challenger* left Sheerness and began its circumnavigation, the Admiralty initially outfitted the expedition with 18 Hydra sounders, then the navy's preferred deep-sea sounding instrument.[11] Thomson described using the Hydra during the short cruise of *Porcupine*, 'where sounding was carried on to the utmost attainable accuracy and at great depths'.[12] Likewise, the Hydrographic Office placed a greater value on readings obtained from the Hydra than from other sounders. In 1871 Richards wrote in his instructions to Nares, 'None of the soundings shown on the chart of the South Atlantic, with the exception of those obtained by Captain Shortland in the "Hydra", can be considered reliable.'[13]

In practice, however, the Hydra could be a contrary machine. Navigating Lieutenant Thomas Henry Tizard related that *Challenger*'s crew had to handle the sounder with care: 'a system was adopted to ease the sinkers down without jerks for almost 400 or 500 fathoms. By letting go of the line suddenly when the sinkers were near the surface, they were found frequently to disengage all at once' and ruined the sounding effort.[14] As well as its habit of releasing the sinkers too soon, Thomson complained that the Hydra only took a small sample of the sea floor. He wrote, 'where dredging is impracticable, and all information as to the condition of the bottom must be got from soundings, some simple adaptation' of a sounder would certainly have a great advantage.[15]

The Invention of 'the Baillie'

The problems of how to survey the ocean floor accurately and to collect materials for observation shaped the context in which the Baillie sounder was invented on board HMS *Sylvia* in 1871. Comparable in size to *Challenger* and launched in 1866, *Sylvia* operated under steam and sail, and regularly conducted deep-

46 Survey ship HMS *Sylvia* pictured in 1870 while deployed in the East China Sea. Sounding innovations made on board *Sylvia* were submitted to the Hydrographic Office and subsequently passed on to *Challenger*, then already on its circumnavigation.

sea soundings. As the Royal Navy's foremost survey ship serving the China station, *Sylvia* undertook detailed hydrographical work around the islands of Japan (image 46). New larger steamships required deeper channels to navigate and the old routes were no longer adequate. To provide a safe passage for such vessels, *Sylvia* used sounders to chart a route through the Inland Sea of Japan. Deep-sea sounding also contributed to building diplomatic relations. Underpinning a British effort to support the Meiji government and modernise the imperial Japanese Navy in the 1870s, *Sylvia* conducted a joint survey with a Japanese naval ship off the coast of Hiroshima.[16]

Part of *Sylvia*'s duties included an exhaustive survey of 'the many groups of islands extending far seaward from the southwest of Korea' and sounding between Japan and China's northern ports. This latter task provided valuable information when laying the first submarine telegraph cable between Nagasaki (Japan) and Shanghai (China), telegraphically connecting Japan to the Asian continent.[17] Although *Sylvia*'s crew successfully deployed Hydra sounders, Navigating Lieutenant Charles William Baillie, who joined *Sylvia* in May 1870 at Hong Kong, took the opportunity to improve its design.[18]

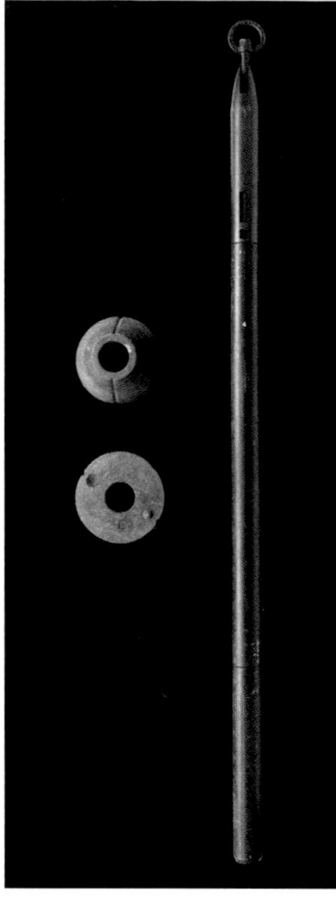

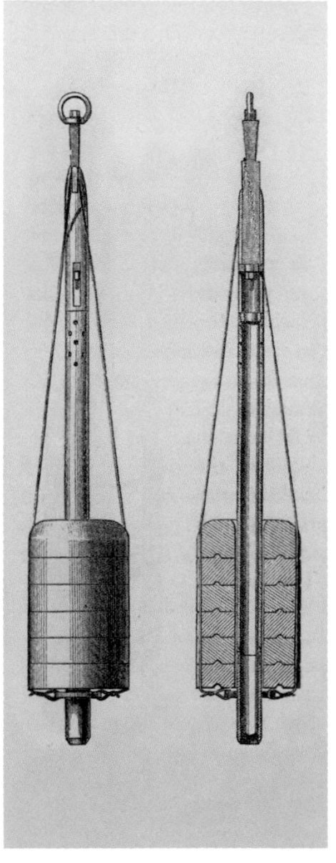

The new instrument closely matched Thomson's desired criteria for a 'simple adaptation' of the Hydra.[19] Besides being more reliable, the device gathered a more substantial sample of the ocean floor, which the 25-year-old officer realised would be a boon to hydrographers and scientists alike.

The 'Baillie', as it quickly became known, was robust but unglamorous (images 47 and 48). Compared to the Hydra's brass and copper sounding tube, its tube was made of iron. The sounder's heavier weight, combined with an increased diameter, provided distinct advantages. Weighing 35 pounds (16 kg), the Baillie sounding tube could withstand the impact of hitting rocks on the sea floor. Moreover, when the Baillie struck a soft seabed, the sounder's weights detached, and the hollow iron pipe continued its downward momentum to pierce the sediments below. This method captured up to 18 inches (45.7 cm) of mud, ooze and sand inside the sounder's tube. The deposits brought to the surface by the Baillie resembled what oceanographers today call a 'core sample', valuable evidence of the geological and biotic composition of the ocean floor. In 1872, after *Sylvia* paid off, Baillie sent a model and design of his new sounder to the Hydrographic Office.[20] Given the problems Tizard noted with the Hydra, a better functioning deep-sea sounder would have been welcome. There, the sounder could be built and replicated, before being mobilised to survey ships for testing and use.

47 (above) The Baillie sounder was used with a hemp sounding line, onto which other instruments could be attached, including a water collecting bottle, pressure gauge and thermometers. In addition to the measurement of depth, temperature measuring was one of the chief objectives of the expedition and readings were used to formulate theories about the global circulation of the oceans and the distribution of marine life.

48 The 'Baillie sounding machine'. The iron cylinder had a butterfly valve at the bottom of the tube and was attached to a brass top; iron weights were held by a wire resting on the shoulder of the top section that moved backwards and forwards freely in the iron tube.

'To prepare as soon as possible a Baillie Sounding Machine'

After receiving a model and design of the 'Baillie', Hydrographer Richards utilised the wide-ranging power of his office to manufacture the sounder and its component parts. Richards was intimately familiar with *Challenger*'s equipment: as part of the expedition's preparations in 1872, the Hydrographic Office had stocked the vessel with everything needed for the voyage, which included an extraordinary range of items, from pencils, notebooks and microscopes, to fishing nets, compasses, chemicals and containers of all sizes. The Admiralty department acquired items from various sources, including private manufacturers, the Royal Society and the Royal Observatory in Greenwich, which was then the navy's centre for testing chronometers. *Challenger*'s deep-sea sounders, however, and the equipment necessary to operate them, were supplied directly from the Hydrographic Office's stores.

Beyond the ship's initial outlay, Richards continued to support the expedition during its voyage around the world. Via letters and telegrams, *Challenger*'s captains regularly reported to the Hydrographer, informing him of their progress, observations and any mishaps or supplies needed. In response, Richards periodically sent materials to the ship. As part of *Challenger*'s ongoing provision, Richards desired to manufacture and send the Baillie sounder to the expedition, then sailing in the North Atlantic. The most favourable place to catch *Challenger* was São Vicente, an island within the Cape Verde archipelago off the west coast of Africa. After this point, the expedition was due to leave the North Atlantic and head further south. If the ship was missed at São Vicente, the next opportunity to send materials was thousands of miles further south, at Cape Town, South Africa.

To build the sounder quickly, Richards called upon the substantial resources of the Royal Navy Dockyards. Expanded from their medieval foundations as places of shipbuilding and repair, by the 1870s many were sprawling industrial and engineering centres. Throughout the nineteenth century, the British Empire depended upon the ships, steam engines and guns manufactured by these facilities. One of these dockyards was located at Chatham on the River Medway in Kent and employed 2,000 artisans and labourers. In 1873, while the Baillie sounder was being prepared, Chatham Dockyard was undergoing an extensive building programme. Between 1862 and 1885, the dockyard added three basins, each covering over 20 acres (8 ha), and four new dry docks. As the Admiralty adjusted to the latest technology of steam-powered ships with metal hulls, private companies increasingly manufactured ship materials, although experimental construction and technological innovation remained the work of naval dockyards.

On 4 July 1873 Richards submitted to William Charles Chamberlain, Captain-Superintendent at Chatham, a request 'to prepare as soon as possible a Baillie Sounding Machine from the Models and drawings sent from the Hydrographic Department'.[21] Baillie had invented an improved deep-sea sounder on board *Sylvia*, but it was the hands of skilled artisans that turned the sounder into a working instrument for use by *Challenger* and the rest of the naval survey fleet. Calling upon the resources of Chatham Dockyard, Richards's order also included an array of sounding gear to accompany the new machines: 310 iron sinkers cut specially to fit the new sounding tube; two iron gin blocks with anti-friction sleeves; 40 deep-sea leads; 70 pounds (31.8 kg) of worsted blue, white and red cloth; 500 iron rings for the sounding rods; and 2,000 yards (1,828 m) of iron wire.[22] In addition, Richards ordered 10,000 fathoms (18,288 m) of deep-sea sounding line to be prepared for *Challenger* at Devonport.

After their construction, the next problem was how to send the Baillie sounders quickly to meet the expedition, then near Madeira off the north-east coast of Africa. Detailed in the department's records, the Baillie and other sounding materials were gathered at the Hydrographic Office and then sent to *Challenger* via a Royal Mail steamship that left Southampton for São Vicente on 9 July 1873.[23] *Challenger* arrived at the island two weeks later, on 27 July 1873, and the crew took the Baillie sounders and other supplies aboard.[24]

The transportation of the sounders from England to the expedition went smoothly, but events at São Vicente proved simultaneously that the transfer of people and materials to *Challenger* did not always go as expected. Along with supplies, the Admiralty sent two men to join the voyage: Sub-Lieutenant Henry Cuthbert Eagles Harston to bolster the officers' work and naval schoolmaster Briant (whose full name is unknown) to instruct Nares's young son. Briant was to replace the ship's first schoolmaster, Adam

Ebbels, who had died suddenly in his sleep (see image 51, p. 92). Travelling on the same mailboat as the equipment and arriving before *Challenger* at São Vicente, Harston and Briant had to find private accommodation. Being a commissioned officer, Harston was invited to stay at the home of the British Consulate. Briant, however, had to fend for himself. With little funds and few provisions, the schoolmaster recorded in his diary, 'I shall be able to manage at least a week if I only sleep at the Hotel. I must pass the time by walking about the island and must do the best I can in the town for food.'[25] When *Challenger* arrived, the schoolmaster was missing. Later, Assistant Steward Joseph Matkin wrote of the man's plight: 'He left the Hotel for a walk, & has not since been heard of. The Consul is of the opinion that he has been murdered, so search parties have been sent out.'[26] He was not found and *Challenger* continued on its course without him.

'It is altogether a new experiment'

The first time the Baillie sounder was deployed on board *Challenger*, the device was a disappointment. The Sounding Log recorded how on Sunday, 24 August 1873, it 'failed to disengage the weights' when it hit the ocean floor.[27] Along the expedition's route south in the mid-Atlantic, from São Vicente, Cape Verde, to St Paul Rocks, Nares reported that the 'Baillie sounding rod is a great improvement over the Hydra, but the hook for the wire of those supplied was turned up and too great an angle'. This defect meant that 'instead of releasing the wire, it nips and jams', and the weights remained attached as the sounder was raised to the surface.[28]

Shipboard innovation characterised the spirit of early ocean research and, soon after the Baillie's failure, Nares altered the instrument's spring mechanism. The hooks that caused the wire to jam were filed off nearly horizontally, so that the lead weights were released as the sounder hit the seabed.[29] He wrote of the Baillie's progress from Simon's Bay, South Africa, on 15 December 1873: 'I can again report most favourably of the Baillie sounding machine; the rod is far more serviceable than the small Hydra-rod, and its weight is not too much for No. 1 line.' For the remainder of the voyage, the Baillie became *Challenger*'s primary instrument for sounding depths below 1,000 fathoms (1,828 m).[30]

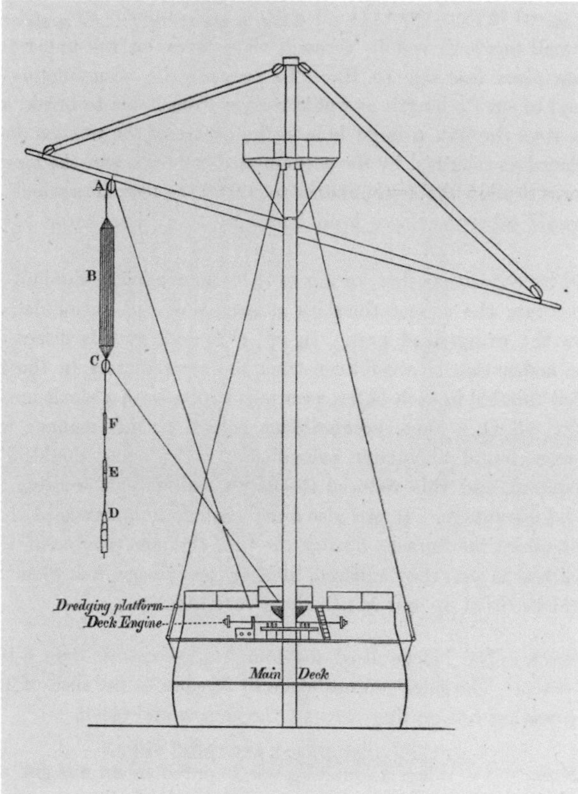

49 Immense lengths of line were led from two large reels on the forward part of the upper deck to a deck engine, then up through a block at the main yard (A) and through accumulators (B) before being attached to the sounding rod (D).

Thomson had taken soundings on the survey paddle steamer *Porcupine*, but deep-sea sounding on board *Challenger* was considerably more complicated. He wrote, 'It is altogether a new experiment to dredge and to take deep-sea observations from so large a ship, and it seems to present some special difficulties, or at all events, to require great management.'[31] Thomson was concerned: 'The roll of the ship, her height above the water, her want of flexibility of movement compared with the vessels which had been previously employed for the purpose, raised new questions as to the method of working.'[32] Part of the issue was that to gain leverage, sounding was conducted from *Challenger*'s main yardarm, a horizontal timber mounted on the mainmast (image 49). As the sounder descended deeper into the ocean, the sounding line, blocks and the yardarm were placed under tremendous strain.[33] The Baillie's sounding weights were at times as heavy as 440 pounds (200 kg); in addition to this were 3,000 fathoms (5,486 m) of hemp line, which in the water weighed roughly 238 pounds (108 kg).[34] The motion of the ship created further tension. To reduce the pressure, 40 'accumulators', massive rubber bands 3 feet (90 cm) long and three-quarters of an inch (1.9 cm) thick, were attached to the line. As the ship rolled, the rubber bands lengthened, absorbing some of the extra tension (image 50).

Despite these safeguards, sounding and dredging in deep water could be hazardous. On 25 March 1873, while on the route from St Thomas to Bermuda, the crew sounded in over 3,000 fathoms (5,486 m). During dredging operations, the pressure on the line was

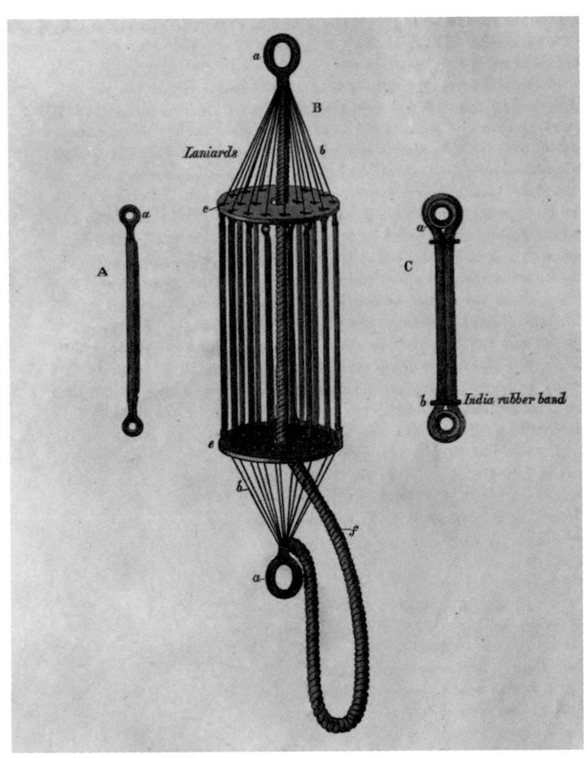

50 Forty accumulators were used for sounding on board HMS *Challenger*, as they were strong enough to withstand the strain of the sinkers on the lead line without being so strong that they gave readily with the motion of the ship. The India-rubber bands were three-quarters of an inch in diameter and could stretch 17 feet.

intense and an iron block broke loose from the upper deck. Matkin related: 'The block as it flew up struck a sailor boy, named Stokes, on the head, & dashed him to the deck with such a terrible force, that his thigh was broken, and spine dreadfully injured.'[35] Despite being attended to by the ship's surgeon, 17-year-old William Stokes died two hours later. The next day, the crew buried the boy at sea (image 51).

Even after repairing the Baillie's flaws, deep-sea experiments remained laborious and time-consuming. From his vantage point as an engineer working in the hold, William Spry described the start of a typical day on board: 'From the hour of four o'clock in the morning, as soon as the watch has been mustered, the bustle and activity begin.' After the crew stowed their hammocks and finished a breakfast of cocoa and biscuit, he wrote, 'The hands dress in the rig of the day, and all preparations are made for sounding and dredging.'[36] *Challenger*'s oceanographic work involved almost every aspect of the ship's operation: a group of officers stood on the sounding platform, stokers worked in the hold, and some 40 sailors handled the ship's equipment and lines.[37] One sounding attempt could take two and a half hours.

At its most basic, deep-sea sounding involved throwing a weight overboard and measuring the length of rope from the ship to the ocean floor, but producing credible data required a much more sophisticated operation. Using sextants, chronometers and computations, navigators ascertained the ship's latitude and longitude so that depth measurements could be plotted accurately on charts. After the vessel and sounder were prepared and their location noted, the officers also recorded the wind force, weather, sea conditions,

51 A memorial erected in Bermuda in 1873 by the crew of *Challenger* to William Stokes, who died hours after a dredging accident, and naval schoolmaster Adam Ebbels, who suffered a cerebral haemorrhage. Photograph by Caleb Newbold, April 1873.

line type, surface current, weights and the machine used: all factors that could disrupt or influence the correct measurement of depth.

The first action was to raise steam, Spry explained: 'Sails are furled, and steam is ready, for it is essential to keep the vessel's head on to the sea during these operations.'[38]

The crew loaded the sounder with iron weights, attached it to the sounding line, hoisted it overboard and released it into the water. As the instrument disappeared beneath the waves, the men worked *Challenger*'s steam engine to keep the vessel directly over the sounding line and to reduce drift, a phenomenon known to exaggerate depth measurements. For instance, in 1853 Lieutenant Commanding Otway Berryman of the sailing ship USS *Dolphin* reported a depth of over 6,500 fathoms (11,887 m) in the mid-Atlantic, a discovery that was celebrated at the time.[39] Two decades later, however, Thomson wrote that 'many of these soundings are now considered untrustworthy' due to errors and inaccuracies caused by drift.[40]

After the sounder's release, the men engaged in sounding closely monitored its descent. Without being able to see the instrument at work, officers used calculations of time, the length of line and the sounder's estimated speed to determine when the device had hit the bottom. The sounding line was tied every 100 fathoms (183 m) with a piece of bright cloth; each time a marker passed into the water, an officer recorded the time elapsed in minutes and seconds. Referring to a table showing the mean rate of descent (image 52), the officers marked when the rate of descent suddenly slowed. This change indicated that the sounder had reached the sea floor.[41]

This system worked well, especially in calm seas, and *Challenger*'s depth soundings were highly accurate, given the technology of the

No. 1 line with 3 cwt. attached						No. 1 line with 4 cwt. attached				
Interval		Time each 100 fathom mark entered water			Depth of fathoms	Time each 100 fathom mark entered water			Interval	
m.	s.	h.	m.	s.		h.	m.	s.	m.	s.
		9	0	0	500	9	0	0		
1	8		1	8	600		0	59	0	59
1	13		2	21	700		2	1	1	2
1	18		3	39	800		3	7	1	6
1	23		5	2	900		4	17	1	10
1	28		6	30	1000		5	31	1	14
1	33		8	3	1100		6	49	1	18
1	37		9	40	1200		8	11	1	22
1	41		11	21	1300		9	37	1	26
1	44		13	5	1400		11	7	1	30
1	47		14	52	1500		12	40	1	33
1	50		16	42	1600		14	16	1	36
1	52		18	34	1700		15	55	1	39
1	54		20	28	1800		17	37	1	42
1	56		22	24	1900		19	22	1	45
1	58		24	22	2000		21	10	1	48
2	1		26	23	2100		23	1	1	51
2	3		28	26	2200		24	54	1	53
2	5		30	31	2300		26	49	1	55
2	7		32	38	2400		28	46	1	57
2	10		34	48	2500		30	45	1	59
2	12		37	0	2600		32	46	2	1
2	14		39	14	2700		34	49	2	3
2	16		41	30	2800		36	54	2	5
2	18		43	48	2900		39	1	2	7
2	20		46	8	3000		41	10	2	9

52 A table showing the mean rate of descent of sounding lines with various weights attached. Based on a chart reproduced in *Narrative of the Cruise*, one of the volumes of the *Challenger Report*.

time. The deepest sounding by *Challenger* was made on 23 March 1875, when a Baillie sounder was deployed south-west of the Mariana Islands, an archipelago of volcanic islands in the western North Pacific Ocean and south-south-east of Japan. The *Challenger* scientists were cautious when it revealed a depth of 4,475 fathoms (8,183 m), but a second sounding confirmed the results. On an 1877 map published by German cartographer Augustus Petermann the location was named 'the Challenger Deep'. While it would take several decades and further voyages for scientists to outline its features, *Challenger* had discovered a flank of the Mariana Trench, the deepest ocean trench on Earth.[42]

'Where soundings are most required'

As *Challenger* continued its voyage around the world, Thomson related that deep-sea sounding was conducted 'at intervals as nearly uniform as possible' throughout the ship's route, a testament to the expedition's scientific mission.[43] A closer look at *Challenger*'s records, however, reveals that his statement does not tell the whole story. Rather than being dispersed evenly, deep-sea sounding was concentrated along points of interest to the Hydrographic Office.[44] Nares's instructions directed that 'independently of the great scientific interest' of exploring the depths, 'it is to be remembered that the rapidly progressing establishment of electric communication between all parts of the earth' required accurate measurements of the 'depths of the ocean, as well as the character and temperature of its bed'.[45] To this end, the captain was 'furnished with a series of charts on which are shown the spots where soundings are most required', including along the routes of future submarine cables between Bermuda and Halifax, and from Sydney to New Zealand.[46] Nearly half of all *Challenger*'s deep-sea soundings were conducted in the Atlantic and closely followed the flow of global trade that connected London with the American continent and the western coast of Africa.[47]

The places during the voyage where no deep-sea soundings were conducted is also telling. From the Cape of Good Hope, *Challenger* travelled as far south as the Antarctic Circle before continuing to Australia, a voyage of 7,637 nautical miles (14,143 km). However, the ship passed south of major trade routes and through a region outside the planned expansion of the telegraph cable network. Although this was the longest *Challenger* spent at sea, the expedition made only

15 deep-sea soundings.[48] Without a clear commercial or navigational motive, deep-sea sounding during this part of the voyage was not a priority. It was also more dangerous to conduct soundings in the Southern Ocean, where the wind and waves were of greater intensity, and coal had to be preserved.

In other areas, the captain and scientists had more leeway regarding where the expedition could undertake experiments. The central Pacific was largely unexplored by deep-sea sounders; in 1870, investors contemplated a telegraphic cable route across the Pacific, but it was not under contract.[49] In this region, Richards indicated, 'it must be left to your own judgment and experience' as to where to sound, as 'all is new ground'.[50]

Marine expeditions, even those that were well planned, sometimes had to adjust to changing circumstances. From Japan, *Challenger* was ordered to 'carry a line of deep soundings across that section of the ocean between it and the coast of America', with Vancouver, Canada, its next port of call.[51] When the American sloop USS *Tuscarora* began a survey of the North Pacific in January 1874, taking a line of deep-sea soundings in preparation for a telegraph cable between the United States, China and Japan, *Challenger* altered its course to avoid covering the same ground. This unforeseen route through the central Pacific led to new breakthroughs, including the discovery of large fields of manganese nodules on the ocean floor. Travelling through the Pacific from Yokohama (Japan) to Juan Fernández (an archipelago off the coast of Chile), *Challenger* spent 116 days at sea and conducted 64 deep-sea soundings, a demanding regime for the crew.[52]

Sounding Results

Soon after *Challenger* completed its voyage, deep-sea sounding underwent radical technological changes, and sounding with thick hemp line became obsolete. The USS *Tuscarora* successfully deployed a sounding machine devised by the British engineer and mathematician Sir William Thomson. Piano wire replaced the heavy rope, making deep-sea soundings more accurate, faster and easier to conduct. The wire sounding machine had been trialled on board *Challenger* in early 1873, but the winding drum collapsed under pressure and the system was deemed too experimental to risk jeopardising the global expedition's results.

Only a year later, while conducting survey cruises from 1874 to 1877, Commander Charles Sigsbee on the US Coast Survey steamer *Blake* further modified William Thomson's sounder. The resulting instrument, the Sigsbee Sounding Machine (image 53), became the trusted model for wire line deep-sea sounding and was used to construct the first modern bathymetric map, a chart of the Gulf of Mexico published in 1888 with over 3,000 soundings.[53] While sounder deployment mechanisms, consisting of the wire line, engine and winding drum on deck, were significantly improved, the Baillie's basic design — a heavy, hollow metal tube with a valve capable of taking a long 'core' sample from the ocean floor — was employed widely for the next 50 years.

Combined with information gathered by contemporary expeditions, *Challenger*'s sounding results began to reveal the hidden topography of Earth's ocean basins. During its time in the Pacific, *Tuscarora*, for instance, made a sounding of 4,655 fathoms (8,513 m) off the coast of Japan, in a region now known as the Kuril Trench and the deepest part of the ocean then discovered, and conducted over 500 ocean-floor soundings and temperature readings.[54] While *Tuscarora* provided a detailed picture of the ocean floor along its route, *Challenger*'s 492 successful deep-sea soundings, following Britain's colonial and commercial interests, occurred across a broader geographical range.

The global scope of *Challenger*'s soundings, enriched by surveys performed by subsequent voyages, changed how scientists and the public imagined the deep sea. The ocean floor beyond the continental shelf was not a uniform, featureless expanse, as many had previously assumed. During the voyage in the Atlantic, Matkin described in a letter to his mother how the depth of the ocean 'is as varied as the land for there are valleys & mountains, hills & plains ... In some places, the dredge brings up mud, & clay, ooze ... in others, it brings up gravel'.[55] Many scientists had thought that a layer of soft, grey mud — made in large part of the shells of dead Foraminifera, a type of single-celled organism — covered the entire ocean floor. Soon after the voyage began, *Challenger*'s scientists were surprised when, at more than 3,000 fathoms (5,486 m), the bottom sediments changed radically in appearance. The 'globigerina ooze' they were familiar with disappeared and in its place they found a deposit they called 'red clay' (image 54), which on inspection contained 'scarcely a trace of organic

53 A Sigsbee Sounding Machine on board the American steamer *Albatross*, 1904.

54 A sample of red clay collected during the expedition.

matter'.[56] In other areas of the ocean, the clay was darker brown or grey, and different 'muds' under the microscope showed the remains of other types of microorganisms.

Among *Challenger*'s more controversial results, but which endured nonetheless, were the expedition's ocean temperature readings (image 55). During deep-sea sounding, the crew often attached other instruments to the line, including deep-sea thermometers and a water bottle, cleverly devised by chemist John Young Buchanan to capture a sample of water at specific depths. William Benjamin Carpenter stressed: 'By the use of these instruments, a vast body of trustworthy information has now been accumulated' better to understand the thermal stratification of the oceans, 'one of the most noteworthy features in the present configuration of our Globe'.[57]

Challenger's temperature readings were not without problems, however. The Miller-Casella thermometers used on the voyage suffered from design defects (image 56). Due to the instrument's small size, measuring only 9 inches (22 cm) in length, the scale's Fahrenheit divisions were impossible to read more precisely than to a quarter of a degree and even this level of specificity was difficult to ascertain. Compounding such errors, the scale and thermometer were not rigidly attached to each other and could move during use. Tizard wrote, 'No instrument of this kind should be sent out of the workshop, to be used on such important work as deep-sea investigation, which has not a scale etched on the stem.'[58] Despite a lack of precision, *Challenger*'s results revealed varying thermal layers and changes in subsurface water temperatures.

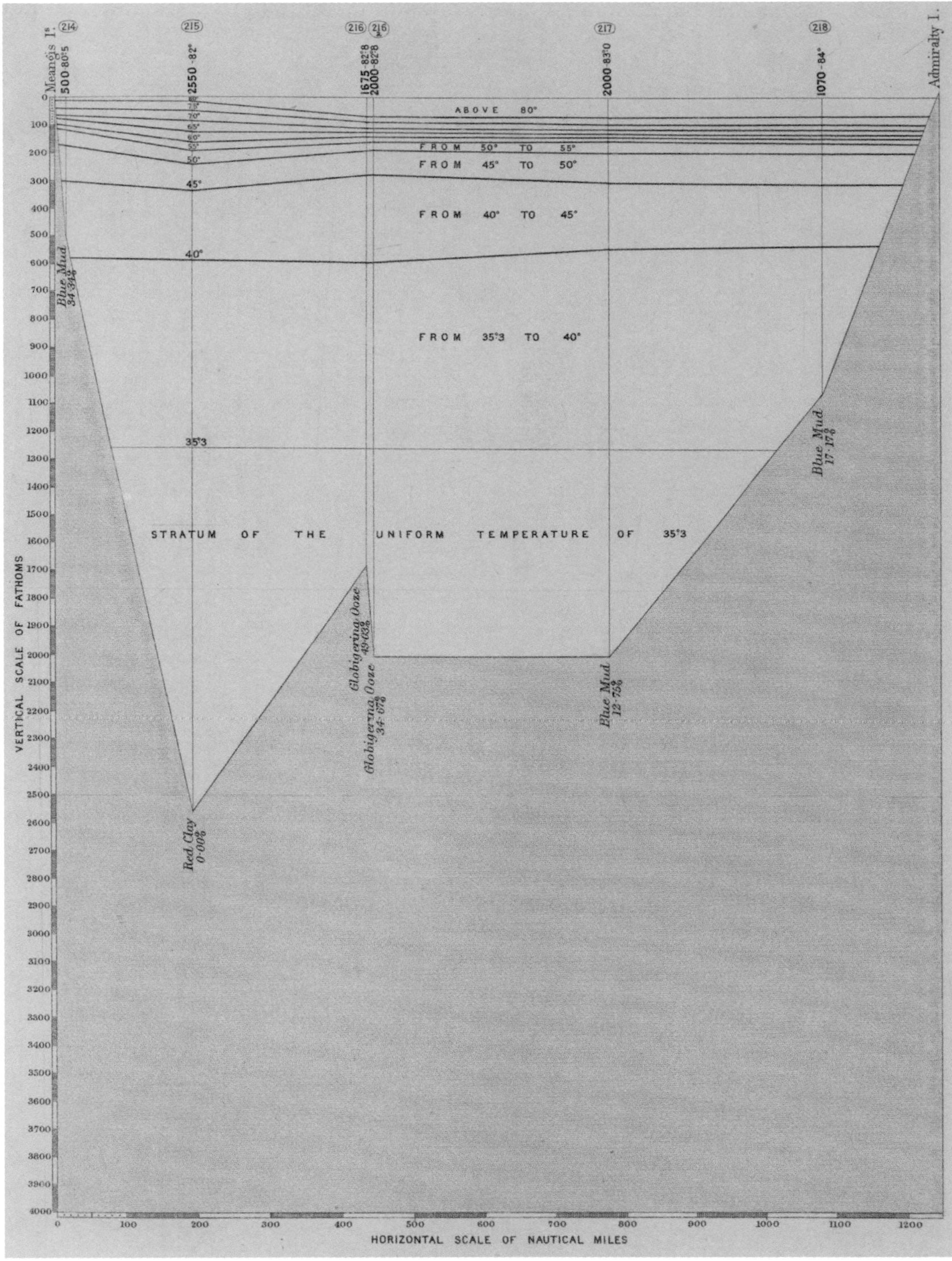

55 Chart of ocean temperatures measured at different depths from Miangis, Indonesia, to the Admiralty Islands in the Pacific Ocean.

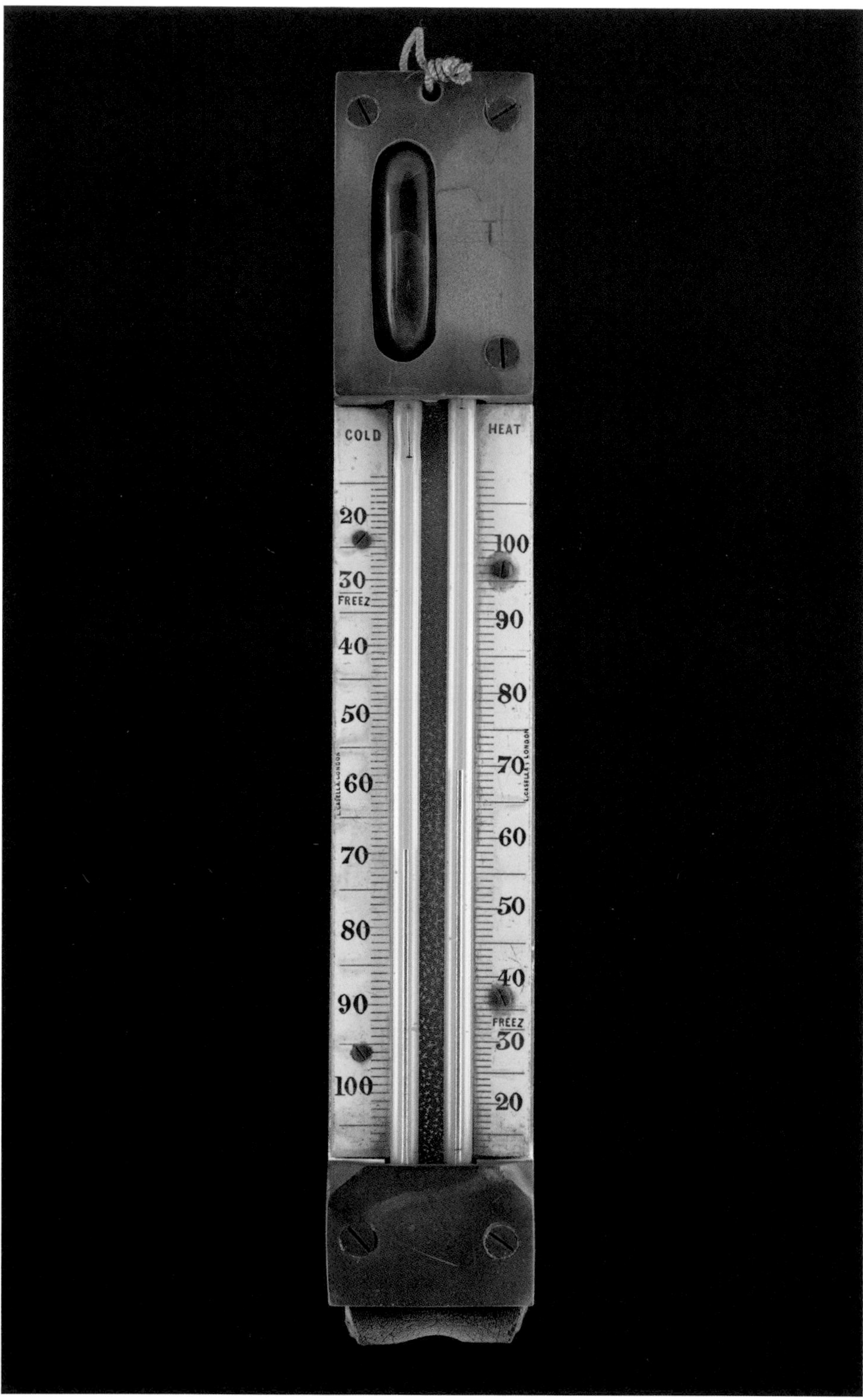

56 Miller-Casella thermometers were used extensively on the expedition. The U-shaped mercury tube recorded maximum and minimum temperatures as the thermometer was lowered into and raised from the ocean.

The mechanics of ocean circulation were being debated fiercely at the time. *Challenger*'s recordings of ocean depth and temperature gave credence to the theory that deep ocean currents are driven primarily by differences in water density, created by variations in temperature and salinity.[59] The expedition's sounding efforts also revealed how underwater features, such as the Mid-Atlantic Ridge, influence the flow and temperature of deep ocean currents (image 57).[60] Adjusted for the systematic errors in the data, *Challenger*'s temperature readings continue to be utilised by oceanographers and climate scientists today, and provide the earliest dataset for long-term observations of global ocean warming.[61]

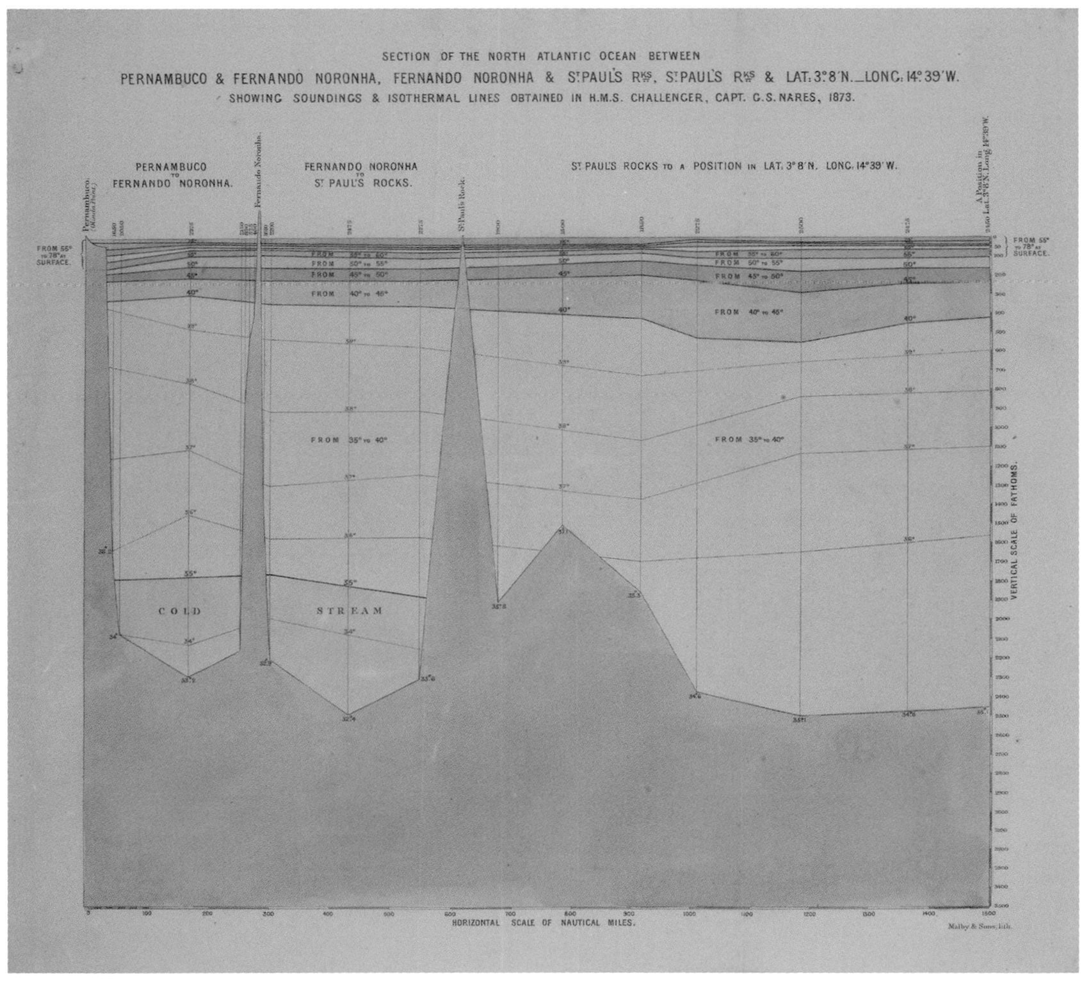

57 Chart of HMS *Challenger* soundings in the North Atlantic, 1873, showing areas of warm and cold water on either side of the Mid-Atlantic Ridge.

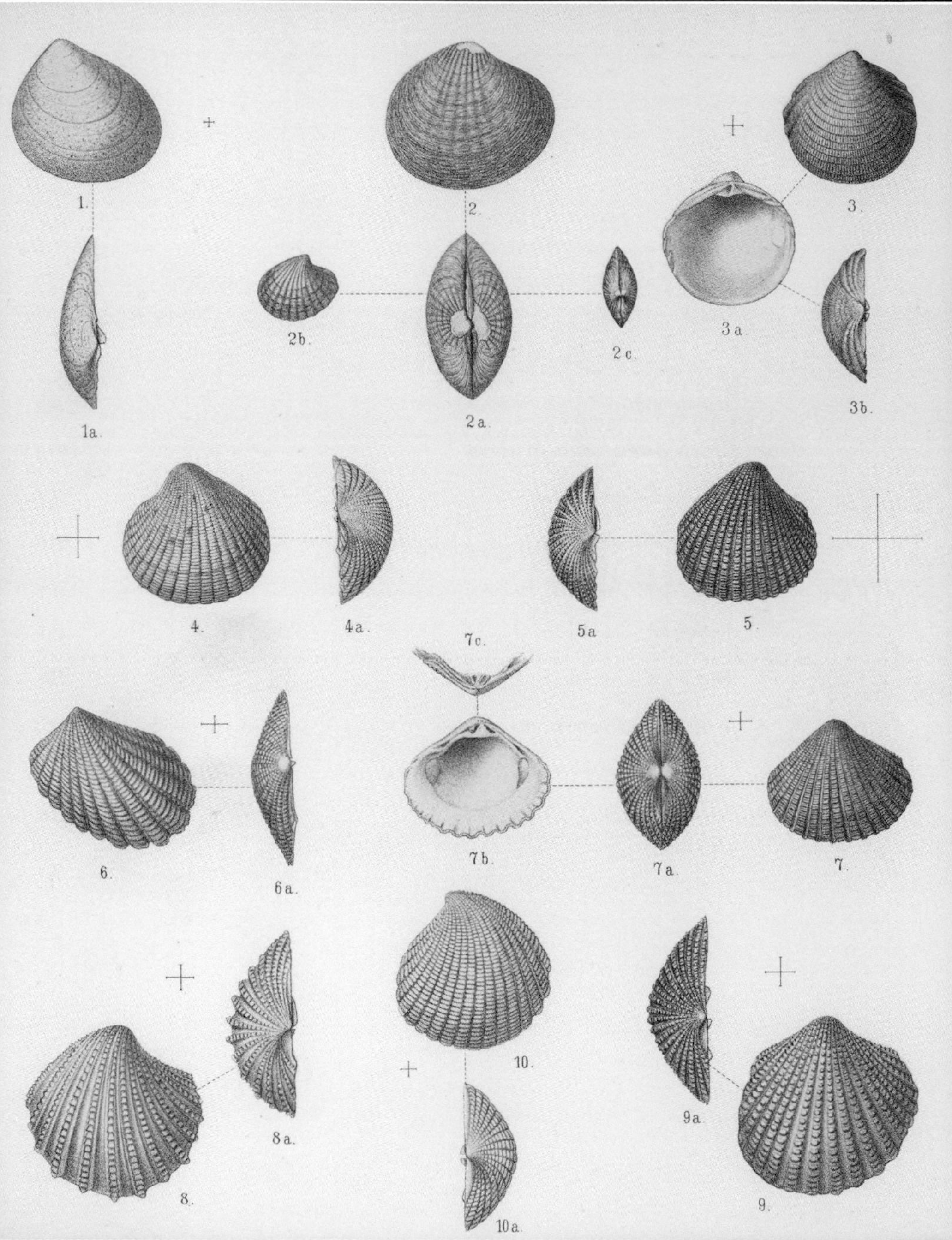

1.
2.
3.
3 a.
1a.
2b.
2 c.
2a.
3b.
4.
4a.
5a
5.
7c.
6.
6a.
7b.
7a.
7.
8.
8a.
10.
10a.
9a.
9.

CHAPTER 4

Collecting Marine Specimens: Finding *C. astartoides*

> We are to visit in succession almost every navigable part of the globe, making a complete circuit of the world and discovering no end of curious and scientific things.
>
> Navigating Sub-Lieutenant Herbert Swire[1]

The quantity and variety of marine animals and plants that the *Challenger* Expedition collected during the voyage from 1872 to 1876 were, by any measure, staggering. Using a dredge or trawl towed along the sea floor, sailors raised to the surface never-before-seen types of worm, squid, octopus and blind fish living in the ocean depths (image 58). Closer to the surface, the crew deployed fine nets to sample plankton, tiny organisms that drift or float in the sea, and fishing line and bait to catch sharks, turtles and fish. Along the shoreline, naturalists combed beaches, investigated coves and explored rivers searching for flora and fauna to collect and preserve. After the ship returned to Britain, Charles Wyville Thomson estimated that the expedition had amassed 600 cases containing over 100,000 zoological specimens for scientific study.[2] The marine animals ranged in size and complexity from microscopic plankton less than 1 millimetre wide to the massive skeletons of sea lions. The travels of a small clam named *Cardita astartoides* demonstrate the remarkable chain of places and people that made the expedition's epic collection and a new era of ocean research possible.[3]

58 Examining the contents of the dredge on board HMS *Challenger*.

Bivalves: Evolutionary Markers of the Sea

As part of its committee's recommendations, the Royal Society asked *Challenger*'s scientists to sample and collect marine plants and animals throughout the voyage (image 59).[4] The discovery of new species remained important, especially the investigation of animals living at great depths. However, scientists also desired to map the general distribution of marine life, a vast global task. By the 1870s, the scientific community generally accepted the theory of evolution, a stance that significantly influenced oceanographic research.

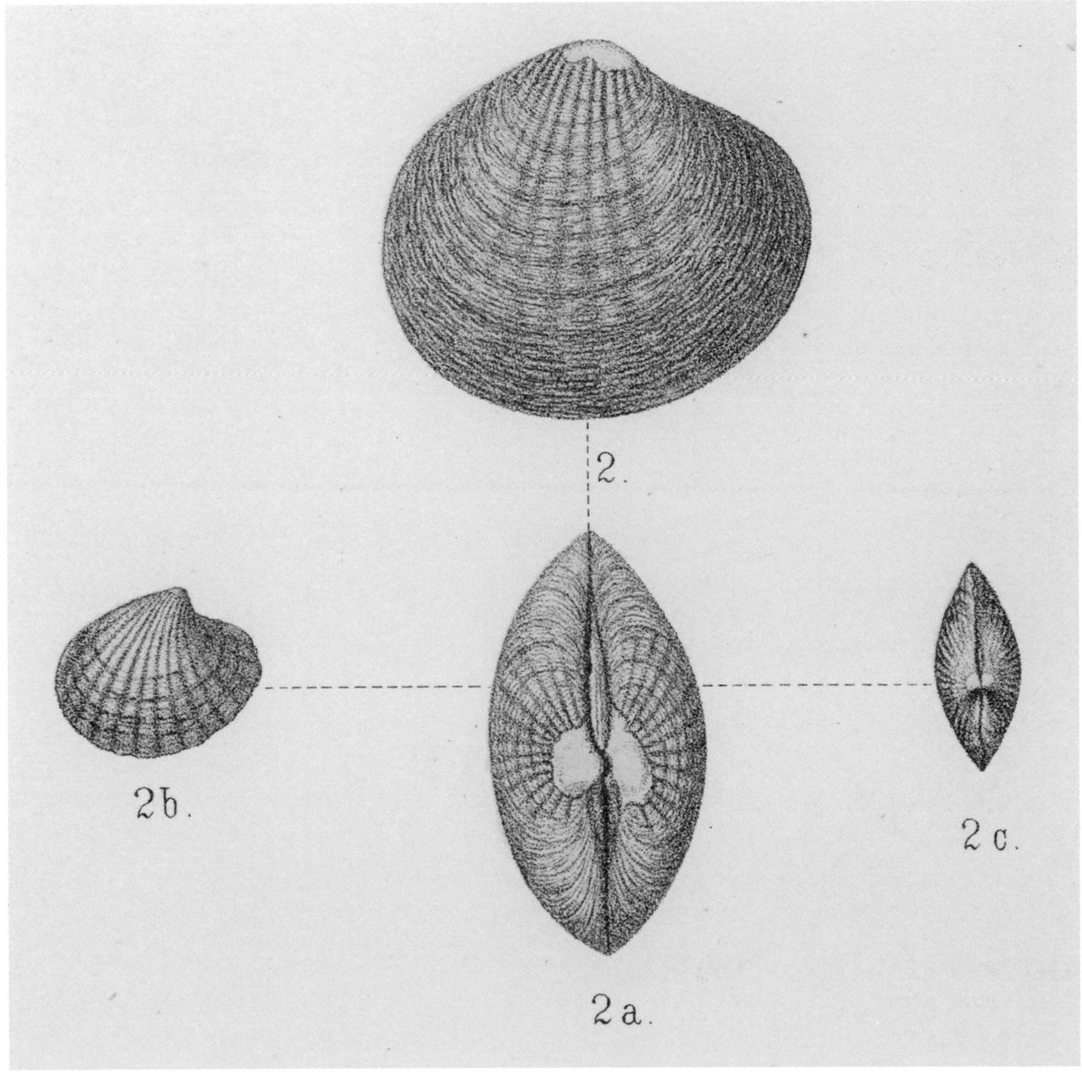

59 Illustration of *Cardita astartoides* from a volume of the *Challenger Report* authored by Edgar A. Smith.

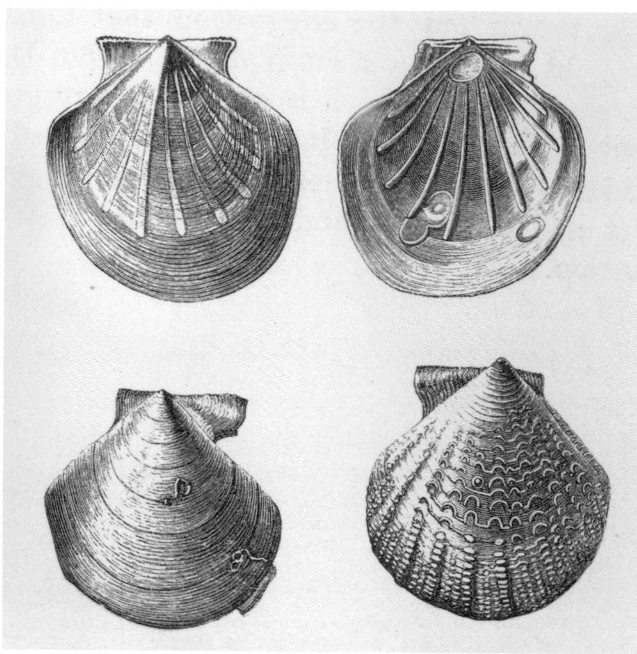

60 Although scientists had little information about deep-sea Mollusca before the *Challenger* Expedition, Professor Thomson considered them a special group.

Knowing where closely related, but differently adapted, species lived in the ocean informed views of changing environmental conditions, migration and evolution. Thus, extensive zoological collections contributed to answering questions of how the ocean had altered or remained the same throughout Earth's geological epochs.

Gathering some groups of marine animals proved more manageable than others. The largest marine phylum, Mollusca, was well suited to scientific collection and study (image 60). (The seven levels of biological classification are kingdom, phylum, class, order, family, genus and species.) This phylum consists of a diverse range of modern and fossil species, including squid, octopus, snails and bivalves, that live in most of the world's ocean regions. Bivalvia (known by *Challenger*'s scientists as Lamellibranchiata), which include clams, oysters, scallops and mussels, are filter feeders that live buried in sediment or attached to rocks. The durable, hinged and often colourful shells of bivalves make these animals attractive to study, appealing to amateur collectors and naturalists alike. Conchology, the study of mollusc shells, had been practised in England since the early 1700s and by the late nineteenth century naturalists knew a great deal about these organisms.

As research questions and physiological comparisons became more complex in the later nineteenth century, scientists desired to study the animals' soft bodies. Acquiring the vulnerable mollusc bodies for research proved challenging, though, especially when collecting specimens from distant places. The United States Exploring Expedition (USEE) obtained molluscs during its oceanographic voyage from 1838 to 1842. Less than six years after the expedition returned, many animals were still unidentified, misplaced or spoiled beyond recognition.[5] Gathering what information he could from the remaining collection, American conchologist Augustus Gould found that individual species of molluscs were not randomly or widely distributed around the globe, as had been thought previously. Although naturalists had observed striking resemblances between shells collected in different regions, Gould discovered that he could identify distinct species by studying the animals' soft body structure. Publishing his findings, Gould recommended that future expeditions preserve soft bodies and shells for prospective study.[6] He insisted that without a soft body to examine, 'we may not consider the question [of species identification] as settled unless the animals have been compared'.[7]

Besides the difficulties of conservation, expeditions had acquired most of the available biological material for research from the northern hemisphere; scientists knew little about molluscs in the tropical zones or the high southern latitudes. The Southern Ocean was one of the least studied areas of the ocean and very few molluscs from the Antarctic were available for comparison and study. The German *Gazelle* Expedition of 1874–76, led by Captain Freiherr von Schleinitz, visited this region in October 1874. Its specimen collection added credence to the theory that marine fauna in the northern polar regions resembled those discovered in the Antarctic.[8] However, hypotheses of global distribution were difficult to confirm with only the few hundred individual specimens that the *Gazelle* was able to procure.

To understand the relationship between bivalves found in disparate ocean regions, scientists needed many more specimens for comparison. Infamous for its high waves, winds and storms, the Southern Ocean proved a formidable place for *Challenger* to add to one of the most significant scientific projects of the age.

Challenger at 'the Land of Desolation', Kerguelen Island

Challenger's surgeon, George Maclean, noted, 'The longest period during which the ship was absent from any inhabited region was in the passage from the Cape of Good Hope to Australia.'[9] From leaving the Cape of Good Hope on 17 December 1873 to the ship's arrival at Melbourne on 17 March 1874, the expedition spent 91 days 'at sea', relying only on its stores and what the crew could obtain from a series of small, mountainous islands along the route. The Royal

Society had ordered that 'special attention should be paid to the botany and zoology' of the islands in the region to the south-east of the Cape of Good Hope. While 'investigations in these latitudes may be difficult; it must be remembered, however, that the marine fauna of these regions is nearly unknown' and 'every addition to our knowledge of it will be of value'.[10] Groups of small islands — named Prince Edward, Marion, Crozet and Heard by previous explorers — were not inhabited continuously but were visited occasionally by American and European commercial whalers, sealers and shipwrecked crews.

Before crossing the Antarctic Circle and investigating icebergs, the expedition explored one of the most isolated places on Earth, the Kerguelen Islands, located more than 2,100 miles (3,300 km) from Madagascar off the eastern coast of Africa in the Southern Ocean. Although fitted with a powerful steam engine and carrying the most up-to-date naval charts, *Challenger*'s voyage in the high latitudes of the Southern Ocean involved significant risk. Hydrographer George Henry Richards cautioned Captain George Strong Nares: 'it is unnecessary, I am sure, for me to impress upon you the extreme caution and vigilance which will be necessary in navigating this boisterous and little known region with a single ship, even in the middle of the summer season'.[11] *Challenger* naturalist John Murray noted that 'the depth of the Southern Ocean and the great extent of open sea over which they blow, enabled the strong westerly winds to produce the longest and highest waves anywhere encountered' during the voyage.[12] The expedition observed swells measuring 18 to 22 feet (5.5–6.7 m) in height from trough to crest during this part of the cruise.[13]

The forces of the seas around Kerguelen made the ship's passage demanding and dangerous; the movement of the vessel from side to side also made life uncomfortable for those on board, severely limiting scientific work. The expedition's official artist, John James Wild, recorded the harrowing experience of travelling through the Southern Ocean on 5 January 1874: 'the ship [was] rolling through an arc of from 20 to 40 degrees each side — 5 and a half rolls per minute. Heavy sea all day.'[14] These conditions prohibited deep-sea dredging and sounding and even writing and reading required much effort. Thomson wryly noted, 'I had been trying to write up some back parts of my journal, but this rolling has I think some bad effect on one's brain'.[15]

Two days later, on 7 January 1874, *Challenger* found refuge from a tumultuous sea. Putting down anchor in a large natural harbour, the crew were confronted by Kerguelen's striking features and mountainous landscape, a wild vista without trees or a permanent human settlement (image 61). Sub-Lieutenant Lord George Campbell described his impression of Christmas Harbour:

> Got in here at nine o'clock this morning — the land of desolation as old Cook called it, and as still called by the sealers. Kerguelen's Land is a gloomy-looking land certainly, with its high, black, fringing cliffs, patches of snow on the higher reaches of the dark-coloured mountains, and a grey sea, fretted with white horses, surrounding it.[16]

Despite its dark countenance and stormy weather, the island provided the crew with sources of fresh food and respite from the turbulent ocean. Ordinary rations at sea consisted of salted beef and pork, preserved vegetables, lime juice, cocoa, tea and biscuit. Here, the men were glad to eat Kerguelen cabbage (*Pringlea antiscorbutica*), an edible wild cabbage found in thick beds on the island's mountainous slopes; they combined it with other ingredients to make soup or ate it

61 *Challenger* at Kerguelen, illustrated by the expedition's official artist, J.J. Wild.

62 Wild placed himself in this illustration of a shooting party on Kerguelen.

as a vegetable. Ducks were plentiful on the island; Navigating Lieutenant Thomas Henry Tizard described how parties went ashore with four or five guns and brought back over 100 birds, a welcome addition to the crew's dinner. Despite the long period at sea, the ship's surgeon noted that the men's health was exceptionally good.[17]

Over the next three weeks, survey parties charted the island's coastline and documented the meteorological conditions. Joseph Matkin wrote, 'parties of Officers & Scientifics went exploring, shooting & botanising inland, but they never succeeded in getting more than about 10 miles' since the 'walking was something frightful, the island one vast swamp. At every other step, you sink up to your knees in the boggy ground'.[18] The parties shot numerous birds and elephant seals for the natural history collections (image 62). At the same time, sailors and scientists collected marine specimens, such as *C. astartoides*, by dredging the waters around the island's coastline.

Knowledge of Kerguelen was not well described on naval charts, but commercial whalers who visited the island periodically had acquired a great deal of information about its anchorages. Once a year, American whaling ships working in the Southern Ocean met with a support vessel at Kerguelen that brought provisions and purchased their valuable cargo of oil and skins.[19] On 27 January 1874, *Challenger* met with Captain Fuller of the American whaling schooner *Roswell King* while anchored in Port Palliser, one of the areas surveyed by Captain James Cook in 1776.[20] Lieutenant Pelham Aldrich described how the expedition 'had been anxiously looking forward to coming across Capt. Fuller who has established quite a reputation among whalers down here, as being thoroughly acquainted with this "Desolation Island"' (image 63). The whaling captain came aboard *Challenger* and guided the ship through uncertain waters into a safe anchorage, which the naval navigators then named

63 White painted wooden graves stood out against the dark landscape of Betsy Cove, Kerguelen. Whalers erected the memorials to their comrades, some buried there and others who perished at sea doing their dangerous work. Illustration by J.J. Wild.

'Fuller's anchorage'.[21] The whalers' intimate knowledge of Kerguelen were thus added to the expedition's findings, as published in charts (image 64), which were used by subsequent voyages.

The Admiralty expected national expeditions to accomplish several tasks at once. One of *Challenger*'s primary aims at Kerguelen was to gather advance information for a British expedition to view the transit of Venus from the island on 9 December 1874.[22] By recording the path of the planet as it passed in front of the Sun from locations around the globe, astronomers could accurately calculate the size of the Solar System. *Challenger*'s officers surveyed the island in preparation to ascertain where the clearest weather could be expected for celestial observations.[23] Navigating Sub-Lieutenant Herbert Swire described a typical day's work of the survey team on 15 January: 'Away all day taking observations for latitude and longitude, and observing angles for a survey of the coast.'[24]

In addition to carrying out surveys and magnetic observations, Captain Nares devoted considerable time and resource to collecting marine life and geologic materials. Given the hydrographical demands of a naval survey ship, dredging was generally regarded as the work of naturalists and rarely practised on warships.[25] Previous naval expeditions had undertaken dredging operations, but captains considered the task a low priority. As part of the USEE, Gould reported that the crew collected few shells offshore, 'very few opportunities having been afforded for obtaining specimens from deep water, by the dredge, on account of the incessant employment of the men and boats on special hydrographical duties'.[26] By contrast,

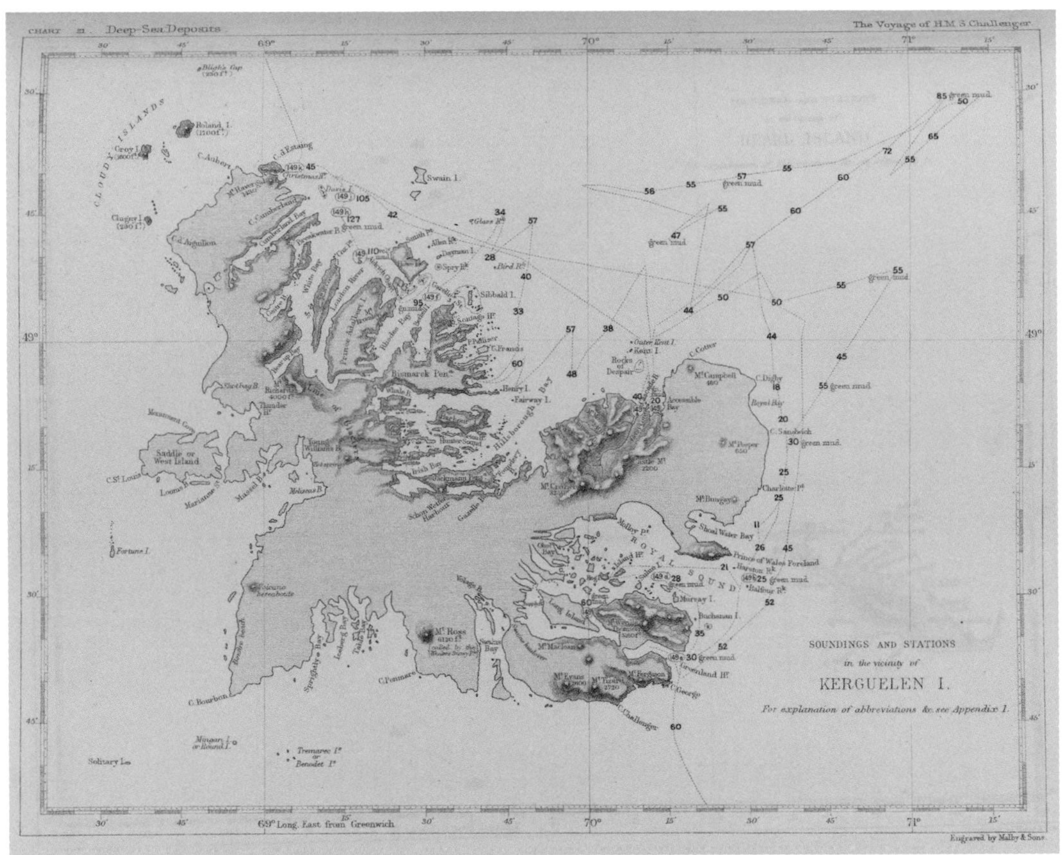

64 Chart showing *Challenger* soundings and stations in the vicinity of Kerguelen Island. Due to dangerously high waves and gales, the island's western side remained relatively unexplored.

from the beginning of *Challenger*'s voyage, Nares fully supported the expedition's scientific programme. In total, the crew performed 111 successful dredgings and 129 trawlings in deep water.[27] More often, dredging occurred in shallower waters with *Challenger*'s tender, a small boat, used to ferry people and materials back and forth between ship and shore. While anchored, the ship's officers continued to conduct surveys, but dredgings were regularly added to the crew's workload too.

Taking advantage of the fine weather on 19 January, a band of *Challenger*'s crew investigated the relatively shallow waters of Royal Sound. The sound was largely uncharted and engineer William Spry characterised the area as a 'labyrinth of islets interspersed over upwards of twenty miles of nearly land-locked waters', a navigation complicated by hazards, such as rocks hidden beneath the surface.[28] To avoid damaging the ship's propeller or hull, *Challenger* remained anchored outside the sound and a group of sailors, officers and

65 *Challenger* carried a small steam pinnace that the crew used for dredging specimens in shallow waters, such as bays and coves. This photograph was taken in Sydney Harbour, by Frederick Hodgeson. *c.*6 April–8 June 1874.

scientists ventured out in the ship's tender (image 65). Thomson wrote: 'The Pinnace was found very valuable for dredging or trawling in shallow water and in a smooth sea.' The boat was 36 feet (11 m) long, with two pairs of small steam engines, one for propulsion and another for heaving in the dredge line.[29]

To gather molluscs and other bottom-dwelling species, sailors lowered a dredge to the sea floor. Then, using the power of the steam engine, the boat pulled the dredge forward, an action that collected materials, including animals such as starfish in the bag's loose hemp strands. Using the second steam engine, the crew of the pinnace raised the dredge and removed its contents. As a result, stated Aldrich, the expedition acquired 'a great quantity of valuable and various specimens' from the island.[30] Dredging also had its drawbacks, however. Material escaped the net during hauling in, especially in deep water, and few bivalves remained intact with the soft parts of their anatomy.[31] It was vital that *Challenger*'s naturalists preserved what soft tissue could be collected for later examination and comparison by specialists.

Preserved and Prepared for Travel on HMS *Challenger*

Killing, sorting and storing marine specimens was a complicated and messy affair; it involved several parts of the ship and much hard work. The three naturalists on board *Challenger* — John Murray, Henry Nottidge Moseley and Ralph von Willemoes-Suhm — probably undertook most of the labour of preservation, but they benefited from the assistance of at least three people, who are rarely mentioned in the expedition's history. Born in Middlesex, Frederick Gordon Pearcey joined *Challenger* at Sheerness at the age of 15 with the rank of Boy 3rd Class. During the voyage he worked as a taxidermist and general scientific assistant. Despite his young age and inexperience, Pearcey quickly learned to identify many marine species; after the voyage returned to Britain he became a member of staff at the *Challenger* Office in Edinburgh. William Pembre joined the ship at Bermuda in 1873, where Thomson hired him as an assistant. He was photographed with the *Challenger* scientists during an outing in the region and also on board *Challenger* (image 66), but died — of 'decline', according to

66 Two assistants to the naturalists, Frederick Pearcey and William Pembre, pose (standing) on *Challenger*'s deck with (sitting left to right) Lieutenant G.R. Bethel, artist J.J. Wild, chemist J.Y. Buchanan and naturalists Henry Moseley and Rudolf von Willemoes-Suhm. Photographer and date unknown.

the ship's doctor — while the ship was at Hong Kong. Little is currently known about another man, whose name was not recorded as part of the official scientific staff. Matkin noted in a letter home that, at Cape Town, Thomson engaged a young Black man to work 'in the analysing room'.[32] Preserving and packing the thousands of marine specimens pulled up by the ship's dredge could not be accomplished by the scientists alone and, besides these men, others may also have been employed in the naturalists' workrooms. After being collected in the dredge, specimens such as *C. astartoides* were brought aboard and the all-important preservation process commenced. Employing a method used by natural philosophers since the seventeenth century, the organic materials were immersed in alcohol to prevent spoilage.[33] Thomson described the practice in detail, noting: 'Our method usually was to plunge the animals direct from the dredge or trawl into spirit of about 84 to 85 per cent.' The naturalists used a strong spirit of wine, highly concentrated ethanol derived from the distillation of wine. Compared to other preserving fluids such as glycerine, Thompson found, 'strong spirit of wine is by far the safest and most convenient medium for preserving marine animals in quantity'.[34] After a few hours, the seawater contained within the animals diluted the solution. Before placing specimens in bottles with fresh alcohol, the naturalists and assistants roughly sorted, examined and entered the general scientific names for each organism in the station book.[35]

Despite these steps, not everything could be preserved successfully. Some deep-sea creatures were too severely damaged by the tremendous pressure and temperature changes experienced while being raised to the surface. Moreover, even if animals were caught and their bodies remained whole, naturalists occasionally discovered rotten fish among the ship's collection.[36]

Keeping accurate records and labelling were vital aspects of the conservation and storage of specimens. In *Challenger*'s station book, the scientists recorded *C. astartoides* as found in Balfour Bay, Station 149, in 20 to 60 fathoms (37—110 m) of water. A series of 504 stations, where the expedition collected specimens and carried out oceanographic experiments, soundings and temperature readings, were marked on charts. For the sake of the scientific project, information such as water depth and the station number had to remain connected to any animals found. 'To prevent any possible confusion, a slip of parchment, with the number written

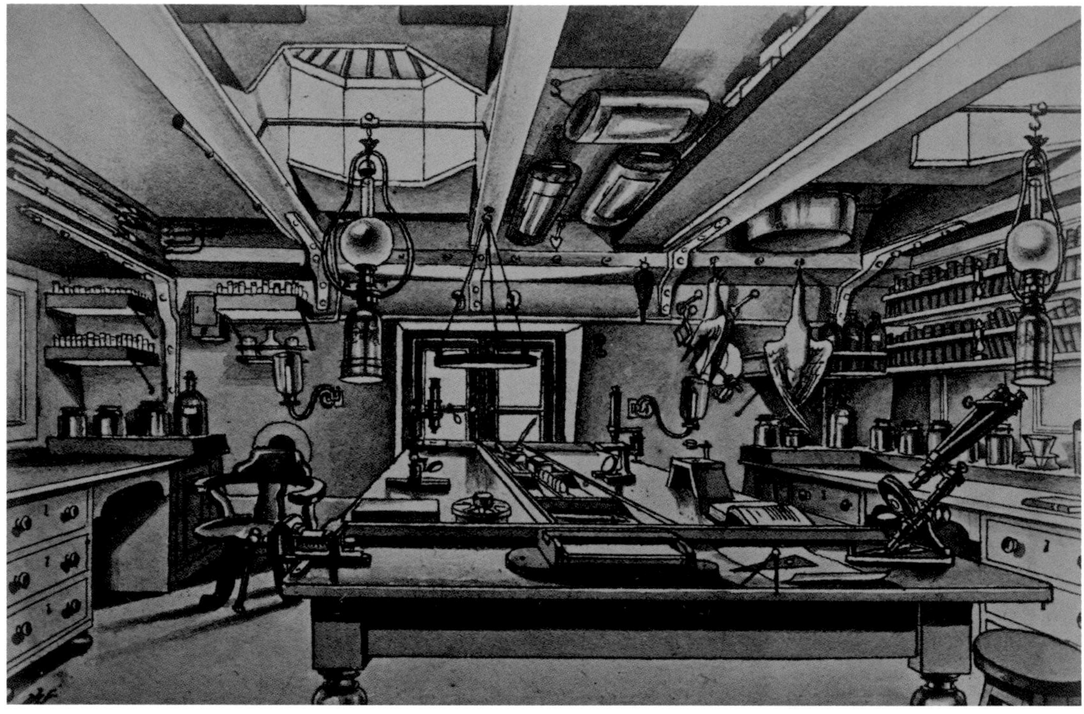

67 The naturalists' analysing room on *Challenger*'s main deck. Instruments such as microscopes were held fast to the table using clamps. Besides a wide array of scientific equipment, the room had an extensive reference library used to help identify specimens.

upon it with a dark pencil, was also dropped into each jar or tin', Thomson wrote. He remarked that the 'duty of filling out the station-books and carrying out these daily details devolved mainly upon Mr Moseley and Dr von Willemoes-Suhm', a task so crucial that *Challenger*'s results were 'due to a great degree to the thorough efficiency with which this duty was performed'.[37] After being labelled and catalogued, the scientific team packed the bottles in wooden shipping crates and sent them to the hold for storage. With meticulous attention to detail, Thomson oversaw the final labelling of the jars and kept the reference books for the contents of the boxes.[38]

Depending on the size of the animals, alcohol was added and exchanged several times to remove excess water and increase the fluid's potency, necessitating a constant supply of spirit. Thomson calculated that each bottle of specimens needed roughly 5 gallons (22.7 l) of alcohol during the preservation process. Upwards of 5,000 bottles and jars were filled during the voyage, meaning the naturalists and their assistants went through an estimated 25,000 gallons (113,652 l) of spirit, equivalent to the volume held in three medium-sized modern tanker trucks.[39] Acquiring this large quantity of alcohol

68 Illustration of the deck-house on board *Challenger*.

was not a problem, however, as it was quickly replenished in principal ports. Sold for the commercial production of solvents or liquors, high-grain alcohol could be purchased in bulk. Thomson noted: 'It was a great advantage to us that we had the spirit duty-free.'[40]

After the sailors collected animals from the dredge, the naturalists and their assistants worked long hours to ready the specimens for transport. During the first part of the voyage, much of the work took place in the natural history workroom on the main deck (image 67). As part of the ship's novel refit, a tank holding about 30 gallons (136 l) of preserving alcohol was installed directly above the cabin and connected by a pipe to a tap in the laboratory.[41] With the fluid available 'on tap', the naturalists could fill bottles with spirit quickly.

Since high-proof alcohol is extremely flammable, most of the ship's supply was stored in the forward magazine, a space in the hold formerly used to house ammunition and explosives (image 69). Keeping alcohol here had two advantages: the cabin was secured under lock and key, thus limiting the crew's access, and it was separated from candle flames, reducing the risk of fire. Placed on racks, 100 metal canisters held 5 gallons (22.7 l) of spirit each. When the laboratory

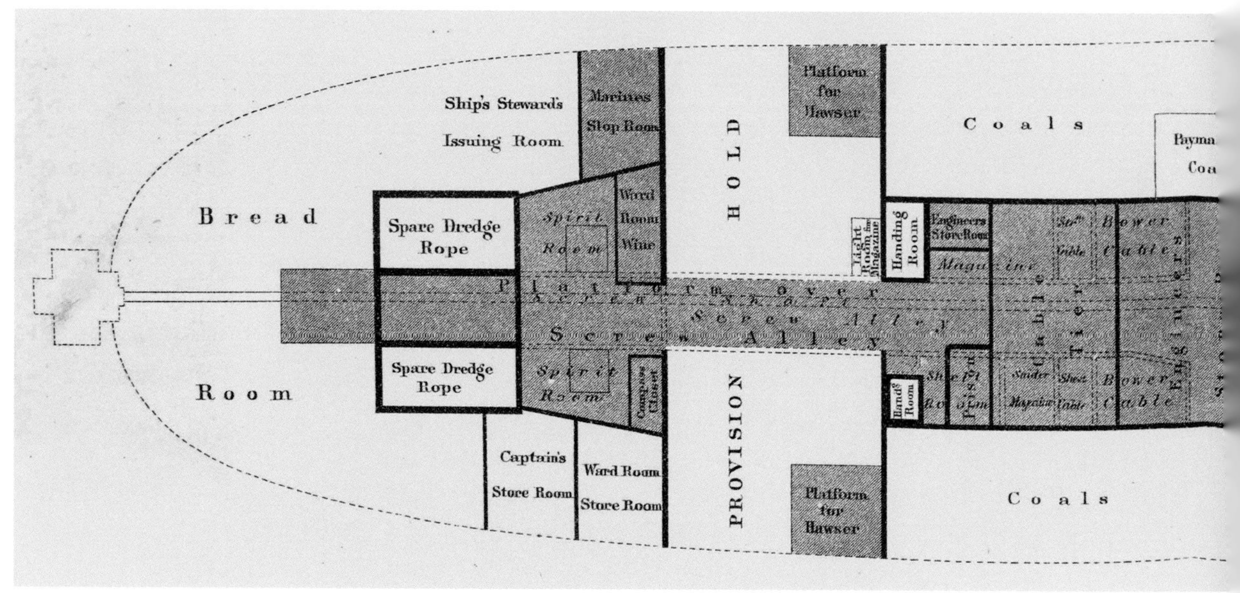

69 Plan of the hold of HMS *Challenger*. Spirits for preserving specimens were stored in a secure area away from flames and the crew.

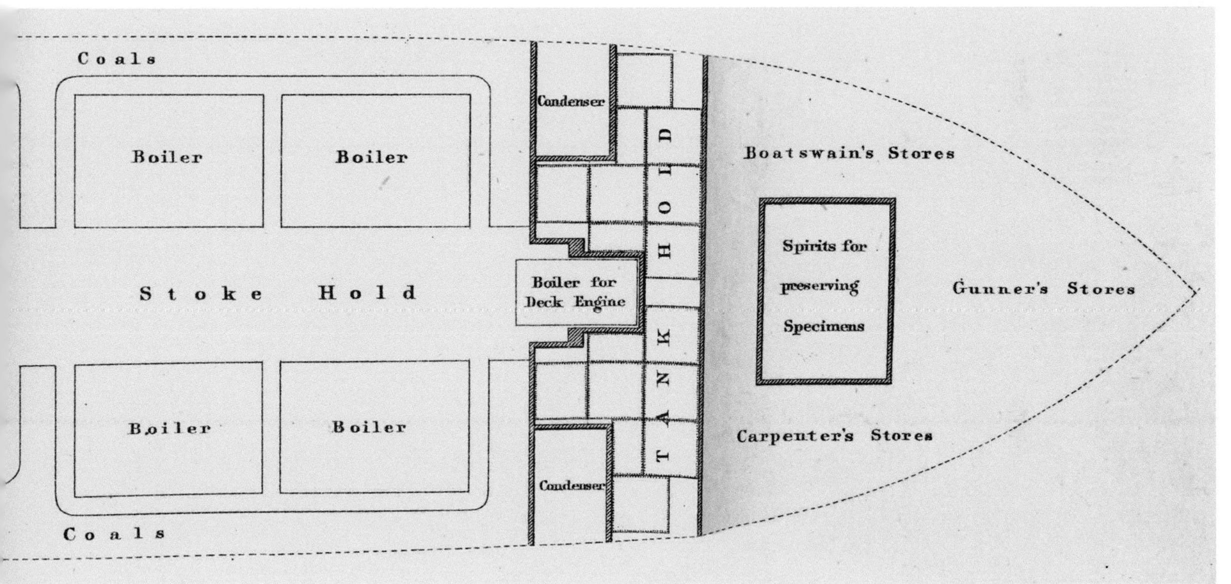
Coals
Boiler
Boiler
Condenser
TANK HOLD
Boatswain's Stores
Stoke Hold
Boiler for Deck Engine
Spirits for preserving Specimens
Gunner's Stores
Boiler
Boiler
Condenser
Carpenter's Stores
Coals

70 The British naval ships *Pearl*, *Dido* and *Challenger* moored at Farm Cove, New South Wales, probably by Frederick Hodgeson, April 1874.

cistern needed refilling, the gunner's mate supervised the careful movement of alcohol from the hold to the main deck.[42]

Sailing in the tropical regions, the foul smells from decomposing animals combined with alcohol, seawater and heat to choke the main deck. Once again adapting and finding solutions, Nares reported: 'More room being required for preparing Natural History specimens, a light deck-house has been built on the after part of the upper deck, abaft the screw well, 8 feet by 6½ feet.'[43] With large windows and plenty of fresh air, noxious fumes wafted behind the ship and not into the captain's cabin (image 68). According to Tizard, the crew also appreciated the new arrangement: 'A deck house such as this, where all the rougher work of the naturalists can be carried on, should be provided in every vessel expressly fitted for researches similar to those carried on in the *Challenger*.'[44]

The work of identifying, labelling and preserving specimens was ongoing throughout the voyage. The bodies of fish, sea urchins, molluscs and other animals kept in alcohol needed regular care and monitoring, which complicated their storage and transport. If too much alcohol evaporated or the solution became diluted over time, specimens were in danger of decomposition. If left unattended for too long, the soft bodies of *C. astartoides* packaged in bottles of preserving fluid would eventually rot.

International Transit from Sydney to Scotland

On 2 February 1874, the *Challenger* Expedition left Kerguelen Island. After a long voyage that included investigating icebergs (see images 72, p. 128, and 89, p. 151) and becoming the first steamship to cross the Antarctic Circle, on 6 April, the ship steamed into Sydney Harbour, an inlet of the Tasman Sea in the South Pacific Ocean. As *Challenger* neared the city, crowds cheered and waved from the shore, and smaller vessels dipped their flags in salute and welcome.[45] The ship first moored at Farm Cove, an anchorage reserved for warships of the Royal Navy and allied nations (image 70). It was a fine anchorage and Aldrich remarked that it placed the ship at the heart of the city's political, social and commercial activity.[46]

The sailors, naturalists and assistants on board *Challenger* had brought *C. astartoides* to Sydney from Kerguelen, but the mollusc still had a long way to travel before arriving safely in Britain. Although the system of preservation generally worked well, the ship's magazine and the naturalists' rooms could not hold the rapidly expanding number of animals acquired wherever *Challenger* went. To solve this dilemma, specimens were occasionally taken off the ship and sent to Britain by other means. Since the 'wet' samples had to be maintained regularly, the time spent in transit was also of consequence and a factor on which the integrity of the collection depended.

Part of oceanography's hidden history is how scientists in the late nineteenth century took advantage of emerging global trade routes to transport natural history materials from around the globe at speed to Europe and the United States. Sydney is a case in point; the city was an ideal place from which to post *Challenger*'s specimens to Britain. With a large harbour and port facilities, it was well connected to London. Spry described it from a mariner's point of view, with its 'beautiful, commodious, and, in fact, the most perfect harbour in the world'.[47] A contemporary guidebook noted, 'The waters of the port are of a depth sufficient for the largest ships afloat to manoeuvre in; whilst as regards its capacity, it is unequalled by any other haven.'[48]

To serve the needs of steamships, Fitzroy Dock, the first sizeable dry dock in the southern hemisphere, had been constructed in 1857.

The Admiralty contributed to the cost of the dock in 1847, 'on condition that it is sufficient for a large frigate or steamer, and that Her Majesty's ships have preference when required for its use'.[49] *Challenger* briefly docked here so that the hull's condition could be evaluated. While Fitzroy Dock benefited British commerce, the colonial project was built by forced labour: some 300 convicts lived on the small island and toiled in appalling conditions for ten years during construction. The prisoners were tasked with excavating from solid rock an enclosed area for the docking and repair of ships. To remain competitive with other ports, during the late nineteenth century the city repeatedly extended the facility to accommodate larger vessels.[50]

In addition to the methods used on the ship, the emergence of global capitalism made it possible for *Challenger*'s specimens to be mobilised quickly from remote sampling sites to Britain. By the 1870s, Fitzroy Dock was only one part of a worldwide system that supported the flow of commerce throughout the British Empire. Engineering feats such as the American transcontinental railway and regular steamship routes allowed people and materials to travel further, faster and more reliably than ever before. In common with the other cities where Thomson unloaded natural history materials from the ship — Halifax, Bermuda, Cape Town and Hong Kong — Sydney was well connected to a number of ports.

The Royal Mail was a vital component of the new, rapid worldwide transport system.[51] Before *Challenger* left Sydney for New Zealand, Thomson posted 65 large boxes and 10 casks of natural history objects to London through the Royal Mail, including a glass bottle containing molluscs of the species *C. astartoides* from Kerguelen.[52] From an examination of Thomson's notes and Australian steamship routes and leaving dates, the *Challenger* boxes and casks were most likely shipped on RMS *Tartar*, chartered by the Australasian and American Mail Steam Ship Company (the A & A Line), which departed Sydney for San Francisco on 6 June 1874.[53]

The steamship route to San Francisco was relatively new; it had been launched in response to the opening of an overland route across the United States, the transcontinental railway. The journey from Sydney to London, via San Francisco and New York, 'was fast becoming a matter-of-course' in the early 1870s.[54] Henry Parkes, Premier of the New South Wales Legislative Assembly, stated that Australia in this system would become 'a mid-way resting-place' in a 'worldwide circuit of communication'.[55] Matkin, who corresponded regularly with his family in England, wrote that mail sent this way arrived about a week faster than his letters sent from Sydney, which travelled through the Suez Canal.[56]

Newspaper stories reveal that the *Tartar*, then carrying 65 large boxes of *Challenger*'s specimens through the Pacific, experienced heavy rain on the way to Honolulu.[57] Unable to obtain noon observations and its precise location, the ship drifted off course. As a result, on 22 June 1874, the steamship struck an uncharted coral reef. The vessel did not take on water, but was grounded for two days and unable to move. The crew and passengers eventually lightened the load by discharging coal overboard; the ship was then freed from the reef and continued to Hawai'i. Arriving at San Francisco docks on 8 July 1874, 32 days after leaving Sydney, *Tartar* anchored off Market Street.[58]

In San Francisco, the Royal Mail shipment was transferred to railway carriages.[59] Railways are rarely mentioned in the history of oceanography but are an essential part of *Challenger*'s story. Completed in 1869, the 'Pacific Railroad' transformed the nation. Travel from New York to San Francisco was reduced from six months to 10 days and to 10 per cent of the previous cost, which was an asset for science as well as business. Collectors took advantage of the railroad to transport natural history materials from the western territories to museums where they were studied, such as the Smithsonian Institution in Washington, DC.[60] Rail carriages took commercial goods and passengers west, and raw materials, buffalo hides, ethnographic materials and newly discovered plants east. Yet, as with docks and coaling stations, the expansion of the railway had consequences for people and for the environment. Along the 1,911-mile (3,075 km) track, Native Americans were forcibly removed from their lands and several thousand Chinese migrants, along with Civil War veterans, performed the dangerous and back-breaking work necessary to build bridges, tunnels and roadbeds.[61]

After arriving at the terminus in New York, the mail was loaded onto a steamer headed to Britain. Boxes containing *C. astartoides* probably arrived at the University of Edinburgh towards the end of July 1874. As Professor of Natural History, Thomson was able to commandeer rooms in which to store materials while the expedition was on its circumnavigation. The university's Professor of Anatomy, William Turner, took responsibility for organising and maintaining

Challenger's specimens until Thomson's return. After the crates were delivered, Turner unpacked the contents and checked the glass jars for breakages or leaks. A large part of his work was to 'have the spirit or brine renewed when necessary, and any reparable injuries repaired'.[62]

The Significance of *Challenger*'s Collection

Challenger's scientists continued the pattern of collecting, preserving, packaging and posting specimens back to Britain throughout the voyage. Due to the preservation techniques used on the ship and the increased speed of global travel, *C. astartoides* and thousands of other animals arrived in Edinburgh intact and well preserved. The system was not without its flaws, though, and objects were lost and damaged along the way. Even with mishaps, by the end of the voyage in 1876 Thomson had acquired an extensive assortment of zoological material, creating what was then the world's most substantial collection of deep-sea animals.

In 1885 London malacologist (a person who studies molluscs) and Assistant Keeper of the Zoological Department at the British Museum Edgar Albert Smith completed his research for the *Report on the Lamellibranchiata Collected by H.M.S. Challenger During the Years 1873–76*, one of the 83 parts of the zoological series on the scientific results of the expedition.[63] By describing and identifying different species of mollusc, he demonstrated that fauna in the deepest parts of the Atlantic and Pacific Oceans were not the same. This had significant consequences for oceanography; it meant that the distribution of marine life was not determined by depth alone. Forces such as evolution and geology must also be at work, a realisation that cast doubt on the theory that the deep oceans had remained unchanged throughout Earth's history.

Several years after the voyage, John Murray returned to the question of why many species from

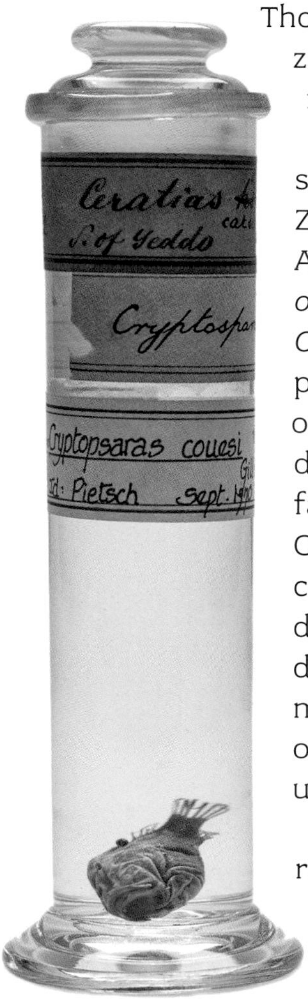

71 A specimen of *Ceratias couesii* collected by the *Challenger* Expedition in waters just south of Tokyo, Japan. The Natural History Museum's Albert Günther originally identified it as *Ceratias carunculatus*, noting its similarity to *C. couesii*, an anglerfish from North America. It was reclassified as the same species almost 100 years later.

the colder waters of the southern hemisphere closely resembled species from similar conditions in the northern hemisphere, but that these species were not found in deep waters in the tropics (image 71).[64] Using *C. astartoides* and other animals from the *Challenger*'s collection as evidence, Murray wrote:

> We are face to face with one of the most remarkable facts in the distribution of organisms on the surface of the globe. A study and comparison of Arctic and Antarctic marine faunas and floras seem indeed to lead directly to very important suggestions as to the past history of the Earth and the gradual evolution to the physical and biological conditions which now prevail over its surface.[65]

He believed that Earth's oceans were dynamic: the oceans' climate has changed over time. Counter to Thomson's previous conclusions, Murray found no indication that their present features, deep water with cold temperatures, had always existed. While explaining the current distribution of marine life, Murray was one of the first to discover that in past geological periods the oceans were warmer and shallower than they are today. The deep oceans were not unchanging relics, untouched by time, as naturalists had suspected. They had their own dynamic history of transformation and upheaval. Thus, the often-overlooked travels of *Challenger*'s collection, including a humble bivalve from Kerguelen, helped Murray and others to bring about a new understanding of the oceans and the history of life on Earth.

21
57 Niagara
58 Boston Mass after the fire
59 I. I. Hayes
May 15 73
60 Musical Festival Boston

CHAPTER 5

Photography at Sea: John Hynes's Photographic Album

> The motion, the dampness of the air, its vitiation by vapours of various kinds, and the extremes of climate that affect the different re-agents and materials — all tell against the photographer.
>
> Charles Wyville Thomson[1]

Challenger is famous for its contribution to oceanography, but the voyage also pushed the limits of expedition photography. French inventor Joseph Nicéphore Niépce produced the first successful photograph of nature, looking out from a window of his home, in 1826.[2] Yet it would take several decades to adapt photographic equipment to travel. For instance, to work outside, photographers had to transport heavy cameras, bulky tripods, noxious chemicals and fragile glass plates. Since negatives had to be developed moments after exposure, even darkrooms had to be made portable. After initial trials and failures, by the late 1860s British army photographers were operating across the British Empire. *Challenger* was unique, however, in that it was the first 'truly global' scientific expedition to embrace photography in an official capacity.[3] Accompanying separate legs of the voyage, Caleb Newbold, Frederick Hodgeson and Jesse Lay captured some 800 photographs, depicting harbours, vistas, geological features, striking fauna, life on board and the rich array of cultures and peoples that the expedition encountered.[4] The men worked in the ship's bespoke photography rooms to replicate hundreds of images for various uses, from newspaper articles to

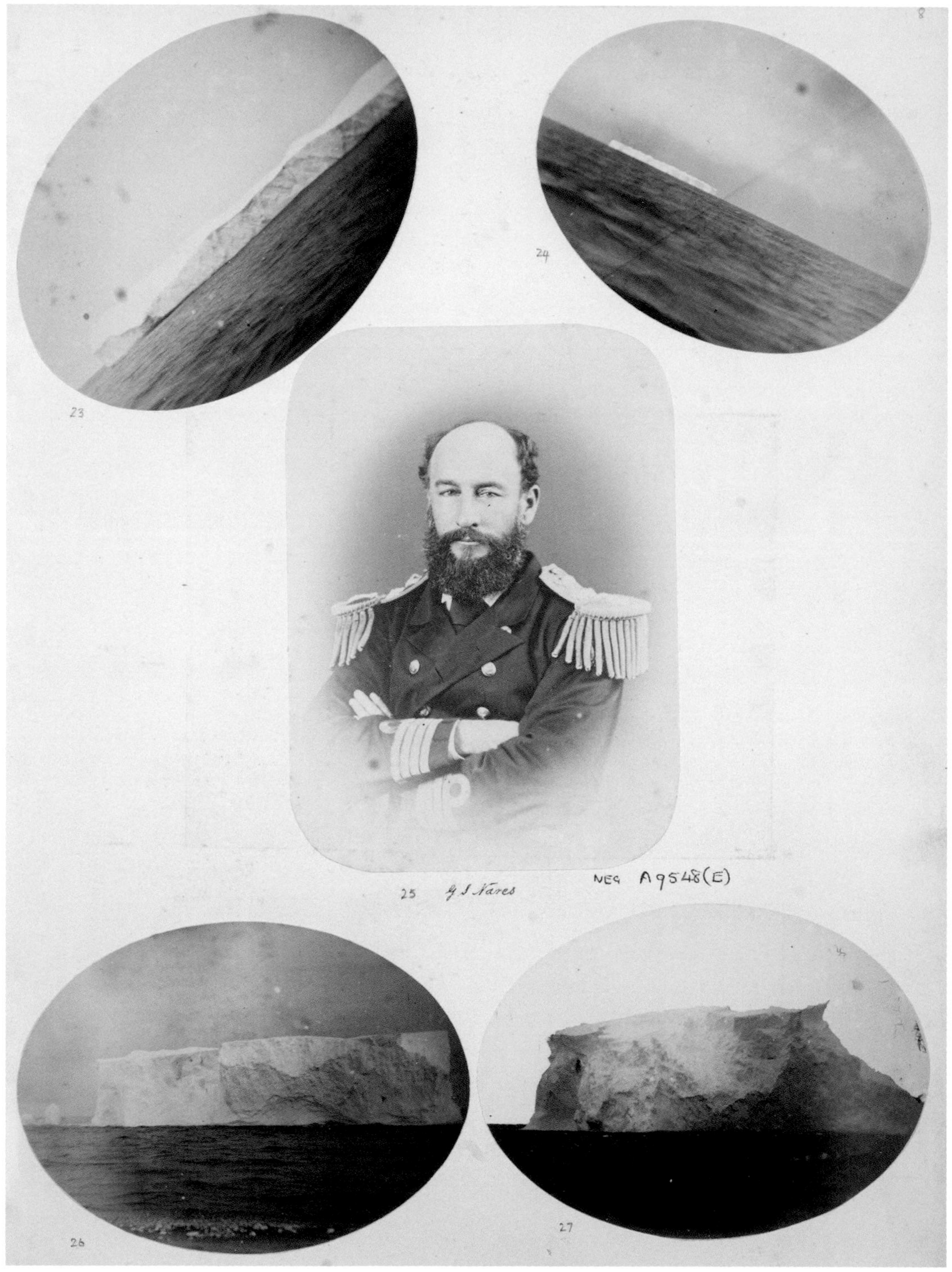

72 A page from one of John Hynes's photographic albums, showing the range of subjects he chose to display. An image of Captain George Nares, likely made after he returned from the British Arctic Expedition in 1876, is included alongside some of the very first photographs of icebergs ever taken. The icebergs were photographed by Frederick Hodgeson between 11 and 25 February 1874, during the expedition's exploration of the Southern Ocean.

scientific illustrations.[5] Individual copies were also sold for a shilling (5 p) each to *Challenger*'s crew members for their personal use.[6] Several officers, including Assistant Paymaster John Hynes, assembled albums from the photographs they acquired on board.[7] Within the pages of the three albums he collated, Hynes incorporated *Challenger* material into his own personal narrative and memories of the voyage (image 72). The photographs reveal glimpses of the men who stood behind the cameras, the places the expedition visited and the people it encountered during its voyage around the world from 1872 to 1876.[8]

The Royal Engineers and *Challenger*'s Photography Workrooms

The Royal Navy played a considerable role in shaping *Challenger*'s legacy, but so too did the Corps of Royal Engineers, soldiers known as 'sappers', who provide engineering and technical support to the British armed forces. Although many of the historical details have been lost, astronomer and chemist Captain William de Wiveleslie Abney, then director of the Royal Engineers School of Photography and Chemistry, outfitted *Challenger* with its photographic workrooms (image 73). Moreover, the first photographer to accompany the expedition, Corporal Caleb Newbold, trained at the school while Abney was its director. These influences connect *Challenger* to the rich history of Royal Engineers photography of which it also forms a part.[9]

73 Sir William de Wiveleslie Abney, by Walter Stoneman for James Russell & Sons, *c.*1916.

In the early 1850s, photography was an appealing technology for use in furthering British imperial campaigns abroad. It was a strategic advantage to possess detailed information about distant places where British troops may be deployed and photography had the potential to record buildings, terrestrial features, fortifications, harbours and roads quickly and accurately. Attempts to photograph the Crimean War, however, revealed some of the major difficulties of expeditionary photography, which the Royal Engineers endeavoured to solve. Not only did photographic equipment need to be made portable, but the collodion process (invented in 1851 by Frederick Scott Archer) consisted of several stages, each of which required dexterity and necessitated working to precise chemical formulas and timings. Beginning in 1856, instruction in photography was offered at the Royal Engineers Establishment at Chatham to increase the photographic skills of the corps.[10] As a result, between the 1850s and 1870s, sappers produced remarkable work, both in the course of their military duties and on assignment to scientific expeditions.

Photographs of faraway places were attractive to a range of British audiences and images produced for military purposes were incorporated into civilian contexts.[11] For instance, in 1868 seven Royal Engineers photographers accompanied the Abyssinia Expedition against Tewodros II, emperor of Ethiopia (image 74). The force was made up of 13,000 British and Indian soldiers, along with 44 elephants, who marched 400 miles (643 km) from the Red Sea to attack Tewodros's fortress at Magdala. The photographic division played a vital role in the coordination of troop movements; they photographed topographical drawings and maps and, by printing

74 The Abyssinian campaign, Ethiopia, 1868. The photograph, taken by a member of the Royal Engineers, shows a military camp at Zoula, with stores and horse lines in the foreground.

multiple copies, supplied some 15,200 prints of plans and views to commanders in the field.[12] The Royal Engineers photographers also recorded landscape and camp scenes, as well as portraits. These images were later assembled into official albums, given to scientific societies and adapted and printed in the *Illustrated London News* to raise public support for British imperial operations abroad.[13]

The expanded use of military photography for scientific and civilian use influenced how the Royal Engineers worked, as well as how they were trained. Abney, a pioneer in photographic theory and practices, took over direction of the Royal Engineers Photographic School in 1871, which was then renamed the School of Photography and Chemistry. Referring to Newbold's participation in the *Challenger* Expedition, he wrote that an 'N.C.O. [non-commissioned officer] whose photographic career I have watched from its infancy for some two and a half years, accompanies the expedition'.[14] As a student, Newbold would have studied Abney's textbook, *Instruction in Photography* (1871). Written specifically for the corps, the book outlined the use of chemical formulas and photographic methods, but also stressed the importance of photography to the fields of science and art: 'Photography cannot be considered merely as a *mechanical* means of reproduction in the present day seeing the valuable aid it gives to almost every branch of industry and art. To become a good photographer, it is necessary to turn to it with an artistic and *scientific* mind.'[15]

As part of Abney's restructuring of the photographic education of the Royal Engineers, he 'built the laboratory and laid down the chemical and photographic system' at the School of Military Engineering at Chatham, an accomplishment that was highly regarded at the time.[16] With his experience and position, Abney was a good candidate to design *Challenger*'s photography workrooms and to procure the range of materials needed for the expedition. He wrote to the *British Journal of Photography*: 'I am glad to inform your correspondent that I had the honour to purchase a very complete photographic equipment for that particular object.'[17] By identifying Abney as the main architect, we gain insight into how the *Challenger* photography laboratory operated and perhaps even how it looked.

Abney was passionate about astronomy and chemistry and, with other members of the scientific community, helped develop new

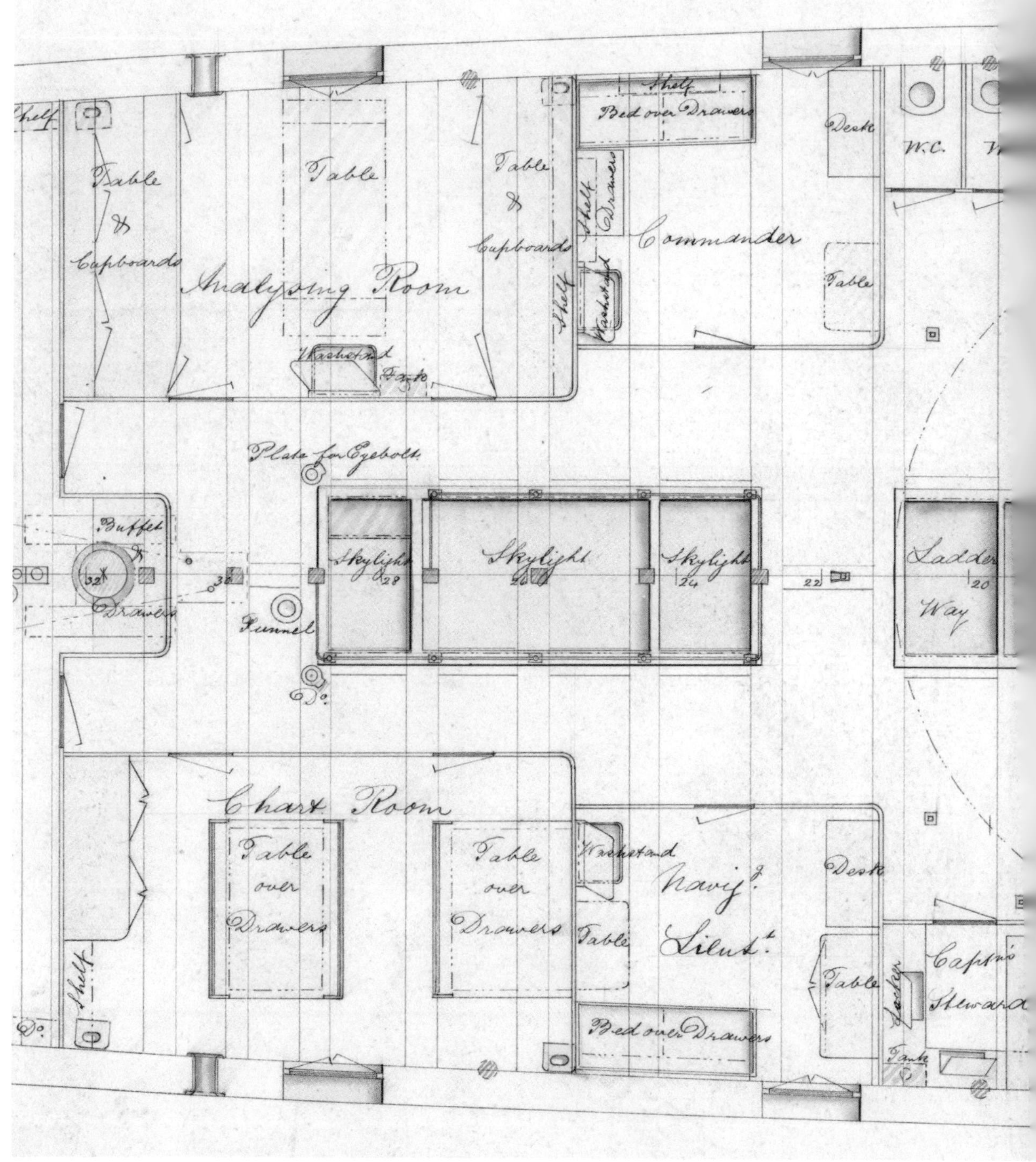

75 Detail from the main deck plan of HMS *Challenger* showing the location of the photographic workrooms and other technical spaces, as well as furnishings and fittings, on board the ship.

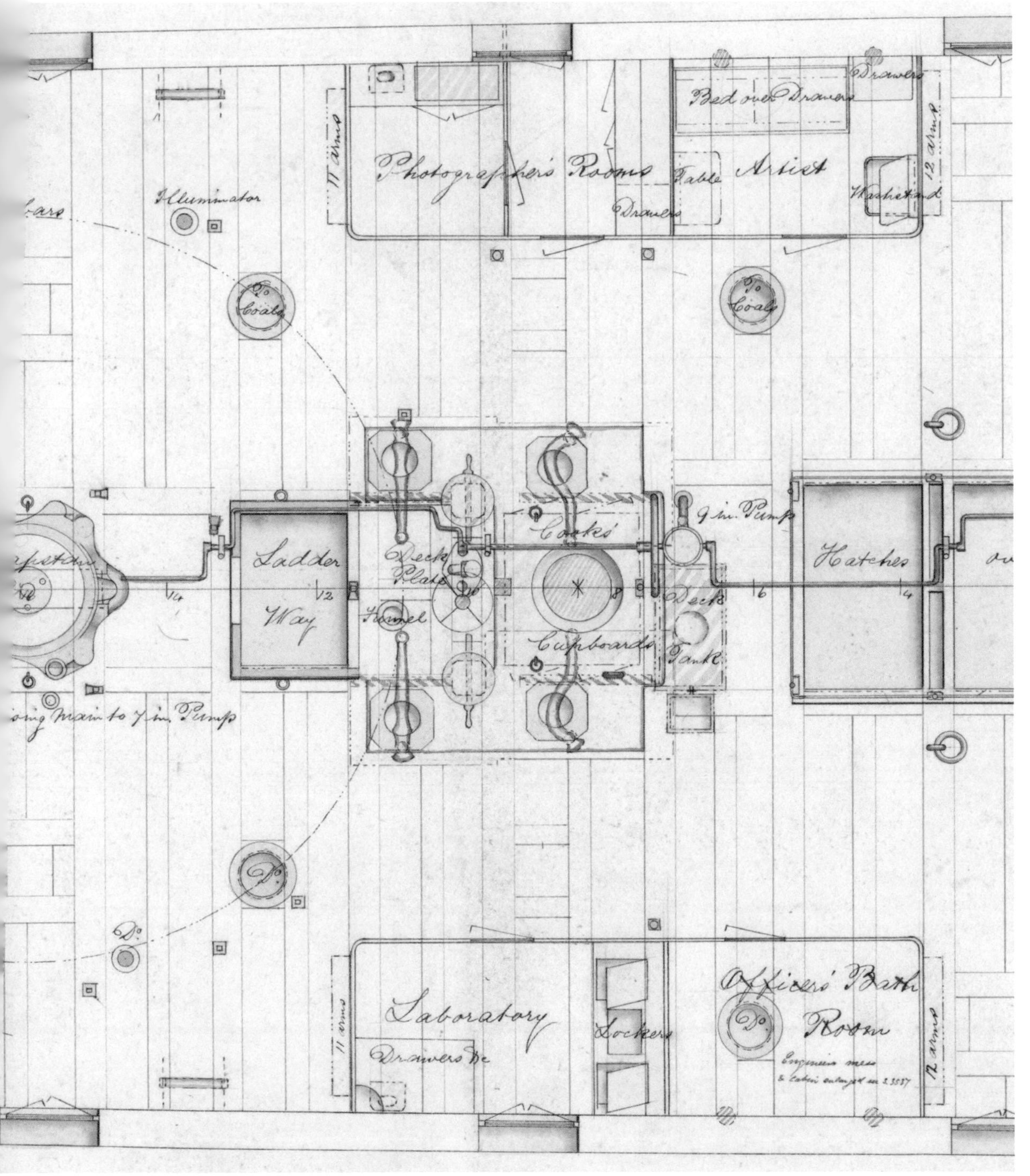
Photographer's Rooms
Bed over Drawers
Drawers
Table
Artist
Washstand
Drawers
Illuminator
Coals
Coals
Ladder
Way
Deck
Plate
Funnel
Cooks'
Cupboards
9 in. Pump
Deck
Tank
Hatches
Laboratory
Drawers &c
Lockers
Officers' Bath
Room

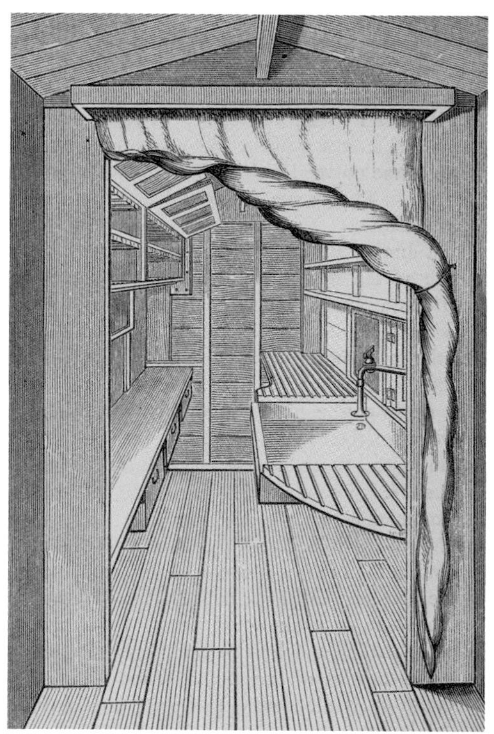

76 The photography workrooms on board *Challenger* may have resembled the portable darkroom used by Abney to photograph the transit of Venus in Egypt in 1875 and shown in this woodcut illustration.

techniques and apparatus for use in the field. For instance, the dimensions of *Challenger*'s photographic workrooms resemble a portable darkroom that Warren De la Rue, a pioneer in astronomical photography, developed for a British solar eclipse expedition to Spain in 1860. De la Rue originally intended to bring with him a 'dark tent', a piece of equipment used by photographers working outdoors. When a large naval vessel, HMS *Himalaya*, became available, the astronomer devised a more substantial wooden structure, which, after arriving in Spain, was moved to the interior by train. Reassembled, one half housed a telescope and the other half a darkroom.[18]

Abney may have modified De la Rue's design for HMS *Challenger*, as he did for his participation in the transit of Venus expedition to Egypt in 1875 (image 76). In describing De la Rue's 'model dark room on a small scale', Abney praised it for its completeness and convenience, claiming it provided all the benefits of a familiar study.[19] From ship plans and diagrams it is clear that *Challenger*'s photography workrooms were constructed in a similar style and consisted of a 'darkroom' and a 'light room', each about 5 feet (1.5 m) wide by 6 feet 6 inches (2 m) long (image 77).

Cameras had accompanied voyages of exploration before *Challenger*. Although never recovered, Sir John Franklin's last voyage to the Arctic in 1845 carried photographic gear.[20] *Challenger*, however, took the practice of photography to an unprecedented level of professionalism. Not only had Newbold received specialist training, but he also had access to purpose-built workrooms with the latest innovations, which allowed the creation and duplication of a great deal of images during the expedition. This enabled photographs to be replicated quickly and adapted for publication in scientific journals as well as the popular press.

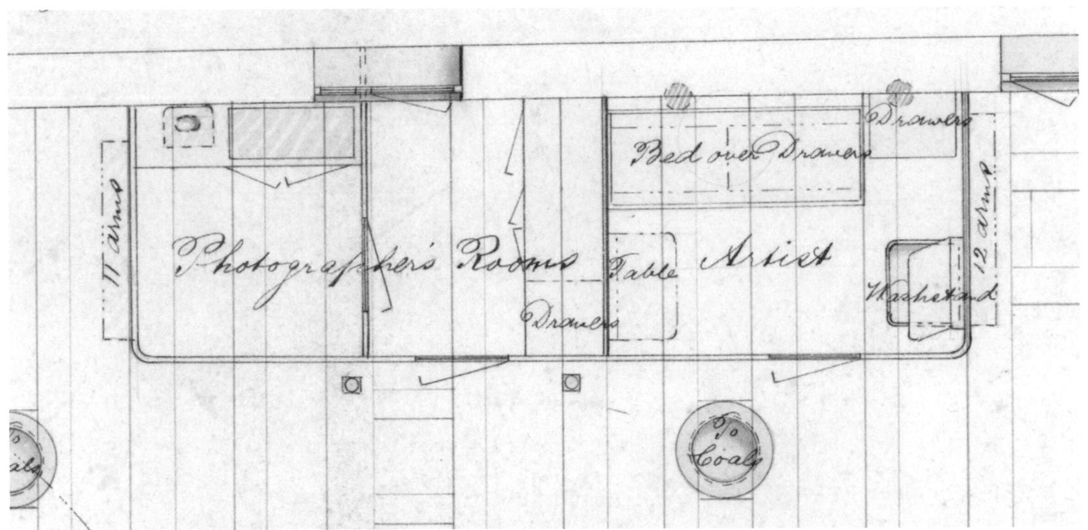

77 *Challenger*'s photographic workrooms were located on the main deck, directly across from the chemical laboratory, and consisted of a darkroom and a light room.

Photography's commercial expansion also impacted on the materials available to *Challenger*'s photographers. In the early 1870s tourists had begun taking photographs while visiting attractions and landscape photographers travelled to remote locations in search of the best views. To serve these interests, companies sold an array of photographic equipment for professional and amateur alike. Abney advised that 'numerous patterns of dark tents were on the market, some very simple and some complicated'.[21] He recommended 'Howard's' tent due to its straightforward construction, ease of use and portability: 'the chemical chest and camera with the attached tent could be easily carried by one person ... it is certainly most convenient'.[22] Equipment such as this, likely to have been included on board *Challenger*, made photographic practice more portable.[23]

Caleb Newbold: A Royal Engineers Photographer Goes to Sea

Abney's textbook, *Instruction in Photography*, provided clear, easy-to-follow instructions. Yet the reality of expedition photography was a complicated and messy affair. Adding to his conundrum, Newbold's position among the ship's crew was an unusual one. Listed as 'Photographer' in the ship's ledger, the army corporal was regarded as a member of the scientific staff, although not a highly paid one. His annual salary was £82, less than half of the naturalists' wages, which gives some indication of his standing within the civilian team.[24]

No official records of Newbold's time on board *Challenger* survive, but the photographs he produced provide clues to his perspective and approach to his work. From the start of the voyage, he created formal group portraits of the expedition's highest-ranking members that exemplified *Challenger*'s dual mission as a scientific and naval expedition. He made his first image when a visiting party from the Royal Society, the scientific gentlemen most responsible for the planning and outfitting of the voyage, posed for a photograph on the deck of the ship before it left Sheerness in December 1872.[25] Another official photograph brings together *Challenger*'s naval staff (image 78); the officers are gathered around Captain George Nares and his young son.

Subtle details reveal Newbold's trained eye and adherence to the 'golden rules' of composition. Abney instructed his students to read the works of Henry Peach Robinson, Vice President of the Royal Photographic Society, who authored *Pictorial Effect in Photography* (1869), which for decades after its publication was the most influential work on photographic practice and aesthetics.[26]

78 Captain George Nares posed with officers and his eldest son, William Grant Nares, who accompanied his father on board *Challenger* at the age of nine, by Caleb Newbold, *c.*1873.

In the image with Nares, for example, the officers' stances are deliberate. On close examination of the sitting officers it is noticeable that each strikes a different pose, echoing Robinson's dictum: 'a row of standing figures, all of the same height, although often to be found in photographs, is eminently monotonous and disagreeable'.[27] Even so, the men are all in harmony with the overall composition. Nares and his son form a study by themselves while remaining the central focus of the group. Objects also add interest to the image: one officer stands holding a sword, while another sits with a telescope on his knee. The placement of the chair on the right, facing backwards, further adds to the variety.

Photographing the natural world, especially the ocean, was less straightforward. How could the ocean, an amorphous and ever-shifting liquid stratum, be depicted via a medium that requires everything to remain still for the best results? Although difficult, it was not impossible to photograph the ocean while at sea. During the first months of the voyage, Newbold experimented with views from the ship, as illustrated by a number of images of Gibraltar (image 79).

Photography, however, failed to capture many other aspects of the ocean that the naturalists studied. The appearance of birds, passing whales, the colours of water and sky, the size and frequency of waves, and the composition of floating debris were recorded not in photographs but in the ship's daily meteorological log and in the naturalists' notebooks.[28]

As the voyage continued in the North Atlantic, Newbold spent many hours in *Challenger*'s photographic laboratory. The room's shelves and cupboards would have been crowded with glass bottles of chemicals, stacks of photographic paper, mixing jars, emulsion trays and glass plates. In this cramped space, preparing glass negatives and developing prints using the collodion process demanded care: the chemicals were poisonous, flammable and potentially explosive. Of *Challenger*'s time in the Atlantic Charles Wyville Thomson stated:

> The management of a photographic studio during a long sea-cruise is a matter of great difficulty. All the circumstances — the motion, the dampness of the air, and its vitiation by vapours of various kinds, and the extremes of climate which affect the different re-agents and materials, all tell against the photographer.[29]

79 Gibraltar, by Caleb Newbold, c.18–26 January 1873.

Not only were the working conditions hazardous, but also the changes in temperature made certain chemical mixtures more unstable and less reliable. Thomson continued, 'Yet in spite of these disadvantages, Newbold has already produced a large number of very satisfactory pictures.'[30]

In February 1873 *Challenger* crossed the Atlantic for the first time. During the 30-day voyage to the Caribbean, the officers and crew conducted 24 deep-sea soundings and 11 trawls, but very few shipboard activities were captured on camera. At anchor, however, Newbold accompanied the scientists and sailors and, following previous British military and scientific expeditions in subject and style, recorded distinctive features of the places they visited. Subsequently, the great majority of his photographs were of topography and landscape scenes (image 80).[31]

Photographs also served as a visual component of the scientists' notes and official reports. The expedition's artist, John James Wild, based many of his drawings on photographs, a technique that allowed him to achieve a level of detail and variety of scene that would have

80 'Group of Palms on the Croquet Lawn, Mount Langton', Bermuda, by Caleb Newbold, *c.*4–21 April 1873, is an example of a photograph of objects of natural-history interest taken by Newbold and the other *Challenger* photographers.

taken him much longer to draw and record initially by hand. Wild's illustrations were then further adapted as woodcuts and printed in the expedition's report and in other books, including his own, *At Anchor: A Narrative of Experiences Afloat and Ashore* (1879) (image 81).[32]

Newbold also set his sights on objects of military and colonial interest, such as harbour facilities, anchorages, roads and buildings.[33] A typical example is a photograph of Hamilton, Bermuda, which chronicles the town's state of development, and reveals a wide street and a row of shops, including 'Darrell's Photograph Rooms' (image 82). Albumenised paper and various chemicals, necessary for printing copies of images from glass negatives, were the materials most often restocked throughout *Challenger*'s voyage. Photographic supplies were replenished at least nine times, either purchased from local merchants or received as shipments from London.[34]

During the expedition's frequent stays on land, colonial administrators and their families often hosted *Challenger*'s civilian staff and arranged for excursions, some of which are shown in Hynes's albums. These events were social as well as scientific;

81 John James Wild, *Challenger*'s official artist, adapted several photographs to be published as illustrations. This image appeared in one of the official scientific reports.

82 Town of Hamilton, Bermuda, by Caleb Newbold, *c.*4–21 April 1873. Aligning with Newbold's artistic sensibilities, the road disappears at an angle to the right and four people pose in the midground, adding scale and interest to the scene.

83 Governor Henry Lefroy (sitting front and left-centre) and Lady Charlotte Lefroy (sitting to Henry's right) shared their scientific knowledge of Bermuda with *Challenger* scientists, including subjects such as the islands' magnetism, botany and geology. Photograph by Caleb Newbold, Bermuda, *c.*4–21 April 1873.

a photograph of 14 people at a 'picnic party' led by General John Henry Lefroy, then Governor of Bermuda, records a visit by *Challenger* scientists to the caverns in the Walsingham area (image 83). In his published memoir of the cruise, *Notes by a Naturalist*, Henry Moseley thanked Lefroy 'both for his kind hospitality and constant information and assistance in scientific matters'.[35] During his time as governor, Lefroy identified and sent a collection of plants to the Royal Botanic Gardens, Kew, which later formed the basis of *Challenger*'s report on botany.[36] From the photograph we know that two women, including the governor's wife, Lady Charlotte Anna Lefroy, attended the intellectual outing. Lady Lefroy was herself a keen observer of natural life and created a series of paintings that illustrated 83 species of plants found on the islands (image 84).[37] Although her husband is given sole credit for acquiring a 'most valuable series of specimens' by the *Challenger* naturalists, it was usual at the time for women's scientific work to be omitted from scientific publications.[38] While her work was not directly acknowledged in *Challenger*'s reports, the photograph provides us

84 Illustration of *Pereskinaculeata* and *Pereskiableo*, Lady Charlotte Lefroy, *c.*1871–77. Lefroy was a keen observer of natural life and created paintings that illustrated 83 species of plants found on the islands.

with direct evidence of her meeting with the ship's scientific team.

The *Challenger* naturalists were intrigued by Bermuda's geology, which they connected to ocean processes. A fortnight of dredging and sounding by the ship had confirmed that the islands' foundations were made of coral, but they found its rocks to be made of another material. Moseley wrote that 'the sand is entirely calcareous, and dazzling white when seen in masses. When examined closely, in small quantities, it is seen to consist of various-sized particles of broken shells'.[39] The sand consolidated to form a type of limestone. When exposed to rain and saltwater it corroded, creating the islands' extensive caves, some of which the naturalists visited with the Lefroys.[40]

The Royal Engineers were successful photographers on land, but some of Newbold's most striking images attempt to depict the great expanse of the open ocean. A series of photographs of the Saint Peter and St Paul Rocks, located about 685 miles (1,100 km) off the coast of north-eastern Brazil, just north of the equator, capture a sense of place in the open Atlantic (image 85). The scientists were struck by the small size of the archipelago: the 15 islets measure less than a third of a mile (0.5 km) from end to end. Thomson wrote, 'we had scarcely realised so mere a speck out in mid-ocean, so far from all other land'.[41]

First discovered by the *Saint Peter*, a Portuguese caravel that crashed against the islets during a dark night in 1511 (and was subsequently rescued by the *Saint Paul*), the rocks had become noteworthy through their distant locale and famous visitors. Captain Robert FitzRoy and Charles Darwin explored the islets in 1832, and James Clark Ross stopped at the rocks during his 1839 voyage to the Antarctic. In homage to these voyages, the *Challenger* crew re-enacted FitzRoy's feat. A boat was sent off and the crew looped

85 Photographic plate of St Paul Rocks in the *Challenger Report*, reproduced from a photograph made by Caleb Newbold during the expedition's visit, 27–29 August 1873.

whale-line around one of the rocks. A hawser, a thick rope or cable used for mooring, was then run from the ship and secured to the whale-line. The ocean current continuously pulled the ship in one direction, holding *Challenger* at a safe distance from the rocks.[42] After being rowed to the rocks and a 'spring and a scramble' jumping ashore at the top of a wave's crest, the naturalists searched for signs of life. They found crabs and insects but no plant life and only a few species of birds, tern and booby, that made simple nests on the larger rocks. Beneath the waterline, however, the islets provided the habitat for an abundance of marine life, which soon attracted the ship's fishing party.[43]

Far removed from the continents, St Paul Rocks presented a stationary point for observing the open ocean. Moseley noted:

> I never properly realised the strength of an oceanic current until I saw the equatorial current running past St. Paul's Rocks. Ordinarily at sea the current of course does not make itself visible in any way ... But St. Paul's is a small fixed point in the midst

> of a great ocean current, which is to be seen rushing past the rocks like a mill-race, and a ship's boat is seen to be baffled in its attempts to pull against the stream.[44]

Although the camera could not capture the deep sea directly, Newbold's photographs of the rocks evoke a sense of *Challenger* exploring a new oceanographic frontier. Due to the cost and complexity of printing, 'St Paul's Rocks' is one of only 19 photographic plates included in the two volumes of the *Narrative of the Cruise of H.M.S. Challenger* (1885).[45]

Photographs also helped to convey *Challenger*'s progress to the British public. Marking the event, Wild created a drawing of the crew mooring to the rocks.[46] His drawing was then photographed on board, and a copy mailed to London at the next port. As with photographs taken during the 1868 Abyssinia Expedition, the image was quickly adapted into a woodcut illustration and subsequently '*Challenger* at anchor at St Paul's Rocks' was published in the *Illustrated London News* (image 86). The article updated the public about the ship's whereabouts and, in the process, created an iconic image that continues to be associated with *Challenger* today.[47]

86 Combining the techniques of photography, illustration and engraving, '*Challenger* at anchor at St Paul's Rocks' appeared in the *Illustrated London News*, LXIII, 1 November 1873.

Leaving St Paul Rocks, *Challenger* crossed the equator into the southern hemisphere at longitude 30° 18' W. As William Spry noted, 'The sparkling light of the North Star had for some time past been growing fainter, and at length disappeared altogether. On the other hand, the Southern Cross, and other stars with which we were not so familiar, had taken their places.'[48] The ship stopped at the archipelago of Fernando de Noronha, site of a Brazilian penal settlement, but a government commandant forbade the *Challenger* naturalists to collect any plant or animal samples. Continuing south along the Brazilian coast, the expedition visited Bahia for two weeks before heading south-east towards the southern tip of Africa. After over a month at sea, the crew arrived at Simon's Bay, Cape of Good Hope, on 28 October. As before, Newbold documented the scientists' excursions on land and recorded topographical features such as an avenue of trees lining a road (image 87), colonial infrastructure like the commodore's house and the diamond mines, a growing source of wealth and interest for investors and international markets. Hynes purchased the first of these images, which was later pasted into his album.

Frederick Hodgeson: Icebergs, Portraiture and Ethnographic Photography

Before *Challenger* left Simon's Bay on 17 December 1873, Newbold made the unexpected decision to abandon ship. It is not known why he left; he may have been daunted by the upcoming voyage to the Southern Ocean or attracted by economic opportunities in South Africa. A replacement photographer was quickly found and a week later Joseph Matkin wrote, 'at the last moment Professor Thomson had to rush up to Cape Town, & engage a young fellow there; who has been very sea sick since coming across'.[49] *Challenger*'s second official photographer, Frederick Hodgeson, appears in the ship's ledger as 'on the civilian staff' and was not listed with the ship's naval complement, meaning that Thomson possibly recruited a professional photographer to take Newbold's place.[50] In the 1870s, it was common for specialised 'harbour photographers' to offer their services to merchant vessels in major port cities and to take photographs of the ship, captain and crew. These photographs were then purchased by crewmen and sent home to loved ones, from

87 Henry Moseley remarked on the shade provided by a grove of pines found along a road in Wynberg, Cape Town, South Africa. Photograph by Caleb Newbold, November 1873.

whom they were sometimes separated for years.[51] Cape Town was a busy port of international trade and Hodgeson may have been a successful harbour photographer convinced by Thomson to join the expedition. Suffering from seasickness, it appears Hodgeson had an uneasy transition to working on a ship at sea.

Like Newbold, *Challenger*'s second photographer was probably trained in classical composition and contemporary methods. In his portrayal of the spectacular and harsh environs of Kerguelen, Hodgeson strategically placed people in the composition, a detail that added a sense of human scale and contrast to the expanse of the wilderness (image 88). It is also an instance when, as Robinson advised in *Pictorial Effect in Photography*, the photographer sought to 'avoid the mean, the base, and the ugly; and aim to elevate his subject' as far as possible.[52] The result of this sensibility is that very few photographs from the expedition show the men doing the arduous or tedious work of the voyage. In a scene overlooking Fuller's Harbour, instead of depicting how a survey group waded up to their knees through wet and boggy moss, Hodgeson photographed a sailor reclining on the ground, projecting a feeling of calm, rather than a struggle against the elements.

88 A sailor overlooking a part of the coastline the officers named Fuller's Harbour, Kerguelen Island, by Frederick Hodgeson, January 1874.

89 'Icebergs in the Southern Ocean, 1874', by Frederick Hodgeson, 21 February 1874. Radiating a sense of exploration, mystery and awe, Hodgeson's photographs of icebergs are some of *Challenger*'s most famous and enduring images.

Leaving Kerguelen behind, *Challenger* travelled south towards the Antarctic. When it came to the problem of how to depict a sense of place in the ocean, Hodgeson again followed Newbold's approach by photographing objects of interest above the water's surface. Early in the morning of 11 February 1874, the first iceberg was sighted, appearing as a long thin rectangle over a calm sea.[53] Sounding, trawling and serial temperatures were taken in 1,260 fathoms (2,304 m), and a few specimens were collected by the trawl.[54] For 15 days, *Challenger* further investigated the ice pack and tested the temperature and the depth of the sea. During the several hours needed to conduct a deep-sea sounding, the ship was relatively stationary, providing opportunities for Hodgeson to take photographs of the ice.[55] Hynes noted in his journal the conditions on board as the temperature dropped: all vapour and moisture froze, and even salt water in a cup became solid.[56] This, as well as the polar light, affected the camera's glass plates, which were covered in a light-sensitive emulsion, as can be seen in the image of a large tabular iceberg in Hynes's album (image 89). Although the harsh conditions resulted in less clear renderings, Hodgeson succeeded in producing the first photographs of icebergs, further advancing the medium as a tool of ocean exploration.

Challenger's scientists were fascinated by the icebergs' complex and unstable structures. Many were flat-topped and bounded by steep cliffs with lines of caves, while others showed signs of weathering, deep channels and toppling over.[57] One of the experiments *Challenger* performed was to shoot a berg with cannonballs to test their effect.

One of the elements that photographs at the time could not record was the magnificent colouring of the southern bergs. The expedition's report related how, at a distance, the general mass appeared to have a slight bluish tint, except where fresh white snow rested on the tops and ledges. On closer inspection, streaks of cobalt blue ran through the ice, with hues ranging from 'common marbled blue soap' to 'the deepest and purest azure blue possible'. On a bright day, the sun illuminated caves and crevasses; one small berg 'looked just like a huge crystal of sulphate of copper, being all intensely blue'.[58] In an attempt to document this phenomenon, several of the scientists and other members of the crew created drawings and watercolours, including a collection made in pen and ink by Able Seaman John J. Arthur (image 90).[59] Even without these colourful elements, however, the *Challenger* photographs of icebergs were widely circulated in scientific reports and newspapers, and Hynes included six in his albums.[60]

As the ship steamed and sailed to Australia and the South Pacific, Hodgeson continued to follow the model of photography established by Newbold through the first leg of the voyage.[61] But the photographs in Hynes's album taken at Tongatapu, the main island of Tonga, illustrate a completely new approach to photography during the voyage. For the first time, the expedition's photographer made a notable effort to create photographic

90 'HMS *Challenger* amongst the ice, Feby. 16th, 1874'. Original artwork by Able Seaman John J. Arthur.

91 (opposite top) Portrait of a man who aided *Challenger*'s safe passage through Tongatapu's coral reefs, by Frederick Hodgeson, 19 July 1874.

portraiture, a medium that sought to express the 'essential' nature of a sitter's identity.[62]

A portrait of a Tongan harbour pilot demonstrates classic elements of Hodgeson's technique (image 91). According to expedition accounts, the pilot arrived in an English whaleboat — a type of wooden boat that could travel under sail or oar, prized by whalers for its speed and valued as a ship's tender — belonging to Siaosi Tupou I, King of Tonga. In the photograph, the man stands alone on *Challenger*'s upper deck and the background is largely blurred. To the pilot's left is the ship's wheel, perhaps referring to his navigational expertise. This type of image is in a style quite different to Newbold's photographs, which depicted people only in groups. After boarding, the pilot guided *Challenger* through the island's surrounding coral reefs to a mooring near the capital city of Nuku'alofa, on the island's north coast (image 92).[63]

The pilot's boat was rowed by four men and overseen by a coxswain, who were also photographed by Hodgeson (image 93). Harbour photographers in the late nineteenth century often produced group portraits in this style. Crew members and officers, a collective force working together to navigate and overcome ocean hazards, were portrayed gathered on deck.[64]

92 The track of *Challenger* through the coral reefs on the northern side of the island of Tongatapu, Tonga.

The photograph of the pilot's gig crew, taken at the request of Moseley, also brings up some of the issues surrounding *Challenger*'s ethnographic photographs. When visiting places of interest, the naturalists made notes on the various landscapes, flora and fauna they found. Like other European and American expeditions, the scientists also recorded and observed the customs, languages and material cultures of Indigenous Peoples. These writings and illustrations were included in expedition reports, scientific papers and other published works.[65]

Moseley, although interested in marine life, believed that the opportunity afforded by the *Challenger* Expedition to study what he and other white Western European and American scientists termed 'the races of men' was of the greatest significance.[66] By 1850, scientists in Europe and the United States overwhelmingly supported the racist theory of polygenesis, the idea that God created each human 'race' as a separate species.[67] Many western intellectuals believed that racial differences, and the alleged superiority of the white race, could be defined by science. This false reasoning supported the interests of the major imperialist powers in the late nineteenth century, which included Britain, France, Germany, Russia and the Netherlands. This was a time when Indigenous Peoples were violently suppressed in Africa, the Americas and Asia. Moseley wrote that some 'races of men' were 'perishing rapidly day by day' and their history needed to be preserved for posterity, a dehumanising attitude, but one shared by his academic colleagues.[68]

Photography was supplementary to the methods by which the *Challenger* scientists studied Indigenous cultures, which also included illustration and acquiring tools, artefacts and human bones for use as scientific specimens.[69] In *Notes by a Naturalist*, Moseley portrayed the language, customs, buildings and physical appearance of people he met during the voyage, with special interest in the people of Te-Moana-Nui-A-Kiwa (Pacific Ocean). Some of this information was repeated in the expedition's official *Narrative of the Cruise of H.M.S. Challenger*, which alongside the scientists' and naval officers' oceanographic findings also collated their ethnographic accounts. Moseley combined his textual descriptions with physical anthropology. On one occasion, he reportedly persuaded one of the crew of Siaosi Tupou I's boat, initially reluctant and suspicious of the naturalist's motives, to give him a lock of his hair.[70] Moseley and the expedition's most appalling actions involved amassing a collection of human skulls and bones, sometimes acquired from colonial naturalists or dug up from graves by *Challenger*'s crew.[71] While Moseley believed his work was critical to record the history of humanity, his dismissal of the people of Te-Moana-Nui-A-Kiwa and Indigenous communities as 'savages' — an attitude prevalent among the *Challenger* staff — was used to justify European colonialism and white supremacy.[72]

Indigenous people were not given much agency in the production of ethnographic images, but, occasionally, they did have some. In the photograph of the Tongan pilot's boat crew posed on *Challenger*'s upper deck, the coxswain wears a pea jacket, a thick garment made of wool. Initially, when Moseley asked him to take the coat off for the photograph — 'in order to make the group uniform' — he refused. Moseley wrote: 'The jacket was a thick garment of the usual pilot cloth, fit only for an English winter, but the man evidently regarded it as a mark of distinction and decoration, and a proof that he was a coxswain.' A deal was made in which the coxswain allowed himself to be photographed without the jacket, separately, only after he was photographed with his jacket on with the group.[73] These photographs and Moseley's recollections demonstrate that, although there was an unequal power dynamic, Indigenous people did have limited say in how they were depicted. In other photos in the *Challenger* collection, women and men turned their heads away from the camera, intentionally making it more difficult for the photographer to capture their likeness.

Challenger photographs of peoples in the South Pacific were used in

93 (opposite) The harbour pilot's gig crew from the island of Tongatapu, Tonga, by Frederick Hodgeson, 19 July 1874.

94 Three Fijian men standing and one sitting on a chair on board *Challenger*, by Frederick Hodgeson, 3–10 August 1874. The canvas-covered hammock storage on top of the ship's bulwarks can be seen in the background.

ethnographic studies, but were also purchased by members of the crew, such as Hynes, whose albums were assembled as a memento of the voyage. While the resulting collections offer little information about the individuals who were not members of the expedition, the images provide glimpses into the lives and communities with which the expedition came into contact from a European perspective. For example, two photographs in Hynes's album depict another boat crew, this time men from Kadavu — Fiji's fourth largest island — who visited *Challenger* (image 94). The expedition was then surveying Galoa Harbour, an important port of call for mail steamers running between Sydney and San Francisco. Besides creating a natural breakwater that protected vessels in all weathers, the harbour's labyrinthine reef formations were of special interest to the naturalists.

From 3 to 10 August 1874, *Challenger*'s small boats surveyed the harbour, while the ship obtained soundings and dredgings off the reef. During this time, the *Narrative of the Cruise* relates that the 'chief at Kandavu [sic] on two occasions sent off a large Green Turtle as a present to Captain Nares'. One of these acts of generosity and

95 A Fijian double-hulled canoe or *drua* that brought gifts from the 'chief of Kadavu' to *Challenger* (then surveying Galoa Harbour, off the island of Kadavu), by Frederick Hodgeson, 3–10 August 1874.

96 *Challenger*'s six Sub-Lieutenants, including Lord Campbell's dog, Sam, probably taken by Frederick Hodgeson between April and November 1874, when he left the ship at Hong Kong. Hodgeson photographed a number of groups on board the ship.

gift-giving was recorded when Hodgeson took a photo from the ship of a Fijian *drua* — a fast, double-hulled sailing vessel — which brought the men and the chief's present to *Challenger* (image 95).[74] The art of building a *drua* is still practised by a few craftspeople and the resulting vessels are regarded as sacred, living objects that connect Fijians with their culture and ancestors.

In addition to portraying Pacific peoples, *Challenger*'s second photographer recorded members of the ship's company, a departure from Newbold's photos, which primarily featured the naval officers and the scientific staff. For instance, a photograph was taken of *Challenger*'s six Sub-Lieutenants, the lowest-ranked commissioned officers on the ship, identifiable in the photo by the single gold ring on their cuffs. Creating a compositional contrast in seniority and duties, they are sitting with Thomson, the Director of the Scientific Staff. The image also depicts a friendly character, one of the pets on board, a dog named Sam, that Lord George Campbell recalled was 'a charming great black curly-haired Newfoundland'. The aristocratic Sub-Lieutenant purchased him from a whaling ship during the expedition's visit to Kerguelen Island (image 96).[75]

Almost exactly a year after *Challenger* left Simon's Bay, the second photographer deserted ship at Hong Kong. Matkin wrote to his mother on 19 December 1874, again updating his family: 'Our Photographer that was engaged at Cape Town left the ship here ... however we have got another.'[76]

Jesse Lay: The 'Missing' Fishermen of Cebu

From his service record, we know that Jesse Lay was born in Winchester, England, and joined the navy at the age of 27 on 2 January 1873. Described as a 'clerk', he was given the rating of Writer 3rd class, and assigned to HMS *Excellent*, a Royal Navy 'stone frigate' or shore establishment, sited on Whale Island near Portsmouth and part of a floating gunnery school. His career progressed quickly: little more than a month later, he was promoted to Writer 2nd class and on 24 November 1873 he joined HMS *Victor Emmanuel*, a screw-propelled 91-gun second-rate ship of the line. Leaving Portsmouth, *Victor Emmanuel* operated as a hospital ship on the Gold Coast of West Africa (now Ghana) during the Third Anglo-Ashanti War (1873–74) before arriving at Hong Kong.[77] There, Lay was discharged on 30 November 1874 and joined *Challenger* the next day as photographer.[78] Unlike the previous two postholders, Lay had experienced the peculiarities of the Royal Navy, which likely made it easier for him to adapt to shipboard life and long periods at sea.

97 Two views of Cebu City, Philippines, by Jesse Lay, 24–26 January 1875.

Lay may have worked as a naval photographer at Portsmouth Dockyard, which was then manufacturing and testing torpedoes, or at the gunnery school, but his training was perhaps not of the exacting standard taught by the Royal Engineers. One difference can be seen in the colour of his photographs: looking closely at Hynes's album, the images from Hong Kong onwards carry a slight golden tint.[79] This could have happened in the course of duplication: yellowing of the emulsion can occur after developing, possibly caused by organic pollution in the water used during the process.[80] Lay also employed experimental 'dry-plates', an emerging technology invented in 1872, in which plates did not have to be developed immediately. Whatever the cause, the small change in colour of Lay's images, made in the ship's photographic laboratory, reminds us that each *Challenger* photographer used methods and chemical formulas that, although similar, were not exactly the same.[82]

98 A specimen of *Euplectella* appears in Hynes's album, a photographic copy of an illustration made by John James Wild. Photographer and date unknown.

Leaving Hong Kong on 6 January 1875, the expedition headed south. As with *Challenger*'s previous photographers, Lay primarily recorded places on land in the manner of a general survey, such as the harbour and buildings on Cebu, the largest island in the Central Visayas region of the Philippines (image 97).[83] However, this type of image offers few clues as to the expedition's oceanographic activities. The waters around the island were of particular interest for the naturalists, Spry explained, as they desired 'to make out, if possible, something of the habits and mode of life of the beautiful sponge, called "Venus's flower-basket"'.[84] The sponge, *Euplectella aspergillum*, was a well-known specimen in nineteenth-century

European natural history museums: when cleaned and dried, the animal's curved and tubular silica skeleton appears as a delicate, white, glass-like latticework about 10 inches (25 cm) long (image 98). Adding to its mystique, cost and rarity, the animals were only known to live in South East Asia, located between the islands of Mactan and Cebu.[85]

Just as they are in *Challenger*'s official reports, many people integral to the scientific project are absent in the photographic record. For instance, there are no known images of the fishermen who acquired the highly sought-after *Euplectella* for *Challenger*'s collection. Rudolf von Willemoes-Suhm noted that the sponges 'were evidently in great supply' in the waters around the island of Cebu, but the British dredges crushed and mangled the fragile forms.[86] The island's fishermen solved this problem by inventing a lightweight instrument that gently pulled the delicate animals from the mud (image 99).[87] Made of split bamboo, the dredge swept an area of the ocean bottom nearly 14 feet (4 m) wide, with 40 to 50 barbs secured to the outer edges.[88] Spry recounted how, 'After about an hour has elapsed, it is hauled in, and several Euplectellas are found entangled amongst the hooks.' Over the course of the six-day visit, he wrote, the 'dredgings were repeated afterwards with great success', and many specimens were obtained, including 'some of very graceful forms, and quite new to science'.[89]

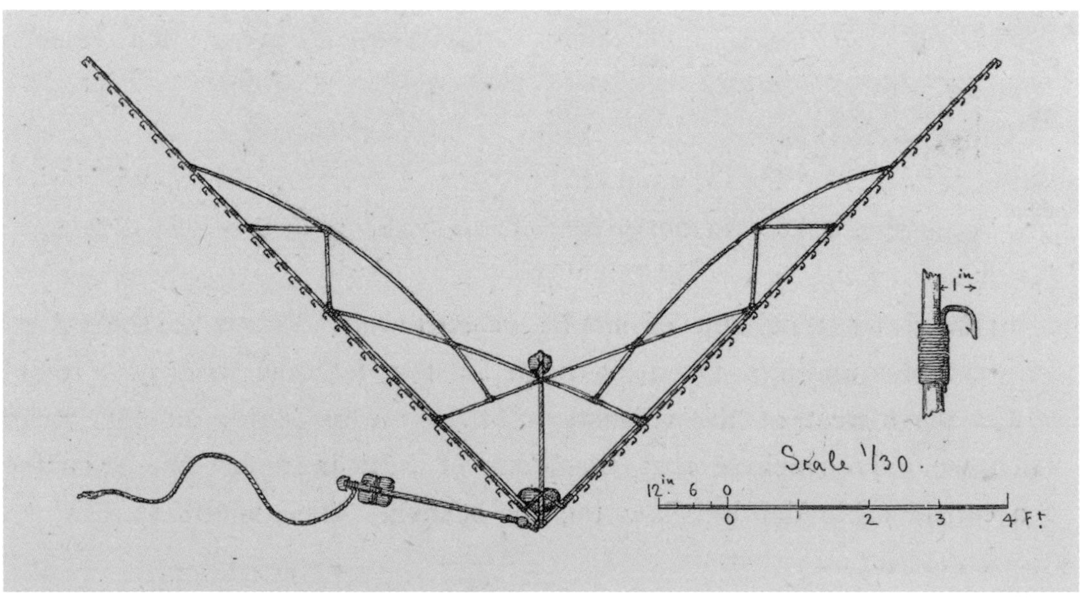

99 'Dredging apparatus used by the fishermen of Cebu'. Although no photographs of the fishermen survive, Wild drew an illustration of their device, which he reproduced in his memoir of the voyage.

As the expedition completed the last leg of the voyage, Lay supported the scientists' ongoing projects of recording natural history features as well as studying the people they encountered. In Hawai'i and elsewhere, Lay took a series of photographs of front and side views of men and women, highlighting their physical attributes (see image 105, p. 166).[90] Made to the same scale for comparison, these images form part of the expedition's ethnographic investigation, whereby people were viewed as racial types rather than recognised as distinct individuals. In this respect, Lay's work, like Moseley's, was typical of prevailing European and American attitudes.

In total, Lay stayed with the ship for 18 months as it travelled throughout the Pacific before passing through the Strait of Magellan and returning to England. In Japan, he produced one of the most striking photographs of HMS *Challenger* (image 100). It is unusual in that it shows the sailors in action, busily working around the ship's hull as it sits in dry dock in Yokohama. Unlike the many other photographs of the warship moored in harbours around the world — which highlighted the strength of the British navy overseas — this image displays the vessel in a more vulnerable state. Propped up by wooden beams, it is undergoing essential maintenance and repair, a potent symbol of the expedition's dependence on the labours and resources of people and places on land to do its oceanographic work at sea.

Expedition Photographs as Personal Memories

When he died in 1926 at the age of 82, Hynes was one of the few remaining survivors of the *Challenger* Expedition. His obituary remarked that he had a 'wide circle of friends' to whom he told 'many a thrilling story of hardship and endurance' from the voyage.[91] Certainly, his albums served as a visual reminder and material evidence of one of the most exciting periods of his life.

Hynes's collection followed a long tradition of naval album-making and storytelling that began with hand-drawn illustrations and personal journals. At some point after the expedition, most likely in the 1880s, Hynes assembled 371 photographs relating to *Challenger* into three albums.[92] Arranged with minimal adornment or modification, the photographs present the circumnavigation in roughly chronological order. Brief, handwritten titles below the

100 HMS *Challenger* in dry dock at the Yokosuka Naval Arsenal, one of the principal shipyards owned and operated by the Imperial Japanese Navy and located south of Yokohama, Japan. It employed more than 2,000 people at the time *Challenger* entered for repairs, including to the ship's rudder, that took seven days to complete. Photograph by Jesse Lay, *c.*3–10 May 1875.

101 HMS *Marlborough* photographed at Naples, Italy, during its service in the Mediterranean (1858–64), before John Hynes joined the navy in 1865. Photographer and date unknown.

102 (below) An unknown woman, probably a close family member of John Hynes, whose image appears on one of the early pages of his albums.

images denote the parts of the world in which they were taken.

Challenger's photographs were shaped by the methods of their respective photographers, the people within them, military and scientific aims, and the varied environments the expedition encountered. Yet in the process of making his albums, Hynes further altered the images by incorporating them within his own personal narrative. In the opening pages of the album, the first ship shown is not *Challenger* but HMS *Marlborough*. Having been born in Ireland, Hynes settled in England as a young man. On 11 June 1865, at the age of 21, he joined the Royal Navy as Assistant Paymaster.[93] HMS *Marlborough* pictured at Naples is a document of his first posting (image 101). On the same page, Hynes pasted portraits of five men and a woman (image 102). The photographs are unlabelled but probably depict members of his family. The contrast between the familiar and domestic ties of home, and the ship moored in a distant Mediterranean port, illustrates that, although the navy offered young men the excitement of travel, they could be separated from wives and loved ones for long periods of time.

Like photography, album-making is a subjective and creative endeavour. Although there are similarities, each of the surviving albums made by members of the expedition is unique. From the wide selection of images produced during the voyage, Hynes purchased and displayed less than half in his albums and excluded hundreds of mundane images – including landscape scenes, buildings and roads — thus obscuring the photographs' origins as a military exercise.

As well as editing the official collection, Hynes added photographs from his personal adventures to *Challenger*'s story. After arriving at the Royal Navy base at Halifax, Canada, on 9 May 1873, the crew worked to prepare the ship for the next leg of its journey. A contingent of *Challenger* naturalists and officers, including Hynes, left the ship and travelled to Boston and New York City in the interim, excursions that were not documented by the photographer. In his album, Hynes recorded his visit to the United States with a collage of seven photographs (presumably bought as souvenirs) depicting Fifth Avenue, Niagara Falls, views of Central Park and a busy street labelled 'Boston'. There is also a signed portrait of Hynes, made by a professional photography studio, dated 15 May 1873, only four days before the ship left Halifax for Bermuda on 19 May (image 103). Within the pages of the album, *Challenger*'s scientific voyage and Hynes's tourist jaunt overlap, shifting the significance of both in the process.

103 At the centre of this page from his first album is John Hynes. His portrait, by an unknown photographer and dated 15 May 1873, is surrounded by tourist images depicting Boston's Great Fire of 1872, Niagara Falls and a musical festival in Boston, US.

104 Two photographs form a view over Bahia, Brazil, with *Challenger* anchored in the distance, by Caleb Newbold, *c.*14–25 September 1873.

Through the placement of prints, Hynes further personalised his storytelling. By joining two separate photographs of Bahia, the viewer sees at once the Atlantic Ocean, the terrestrial landscape and the built environment of the city — a sweeping vista that the photographer could not capture through a single image alone (image 104).

In making his album, Hynes also altered and combined ethnographic studies into a narrative of the voyage influenced by his personal experiences. He trimmed Lay's depictions of a man and woman from Hawai'i into oval shapes, and, in doing so, made them appear more in the style of studio portraits. He then affixed the photographs around an image of King Kalākana of the Kingdom of Hawai'i, who visited the ship during its stay in Honolulu between 27 July and 1 August 1875. When shown a specimen under the microscope, Kalākana recognised 'the well-known anchors in the skin of the Holothurian *Synapt* [a sea cucumber]' and 'named them at first glance'.[94] As seen in a photograph taken by Lay documenting the event, a carpet was rolled out on the ship's deck and both Captain Frank Tourle Thomson (who replaced Nares at Hong Kong) and Charles Wyville Thomson hold ceremonial swords, marking a royal visit that Hynes would have witnessed even if he is absent from the picture (image 105). Viewed together in Hynes's album, these images demonstrate how expedition photographs created for scientific and diplomatic reasons were adapted and integrated into other formats, from the mass media of newspapers to the personal format of photograph albums, and gained additional layers of meaning in the process.

105 King Kalākana, pictured here on *Challenger*'s deck (centre), promoted education during his reign and was knowledgeable about deep-sea discoveries. The identities of the man and woman are unknown but it is likely they lived on the island of O'ahu, Hawai'i. Photographs by Jesse Lay, 27 July–4 August 1875.

Photographs depicting *Challenger*'s voyage continue to circulate widely and can be viewed online and in books written about the expedition. Despite this, many of the images, scattered throughout museum collections and private holdings, have not been studied or catalogued to any great extent.[95] Information about the individuals in them, other than the scientists or naval officers, was not well preserved, presumably because it was deemed unimportant at the time. The prioritising of western involvement and beliefs over that of the communities encountered was typical of colonial attitudes to exploration and discovery. The knowledge that Indigenous Peoples have must be respected and its inclusion in the history of oceanography and *Challenger*'s story is long overdue. Regardless of this omission in the historical record, no matter where *Challenger* travelled, the expedition benefited from local networks and expertise in creating its global study of the ocean.[96] A more comprehensive history should include the lives and experiences of the communities the expedition visited, even if they are currently anonymous in *Challenger*'s photographs.[97]

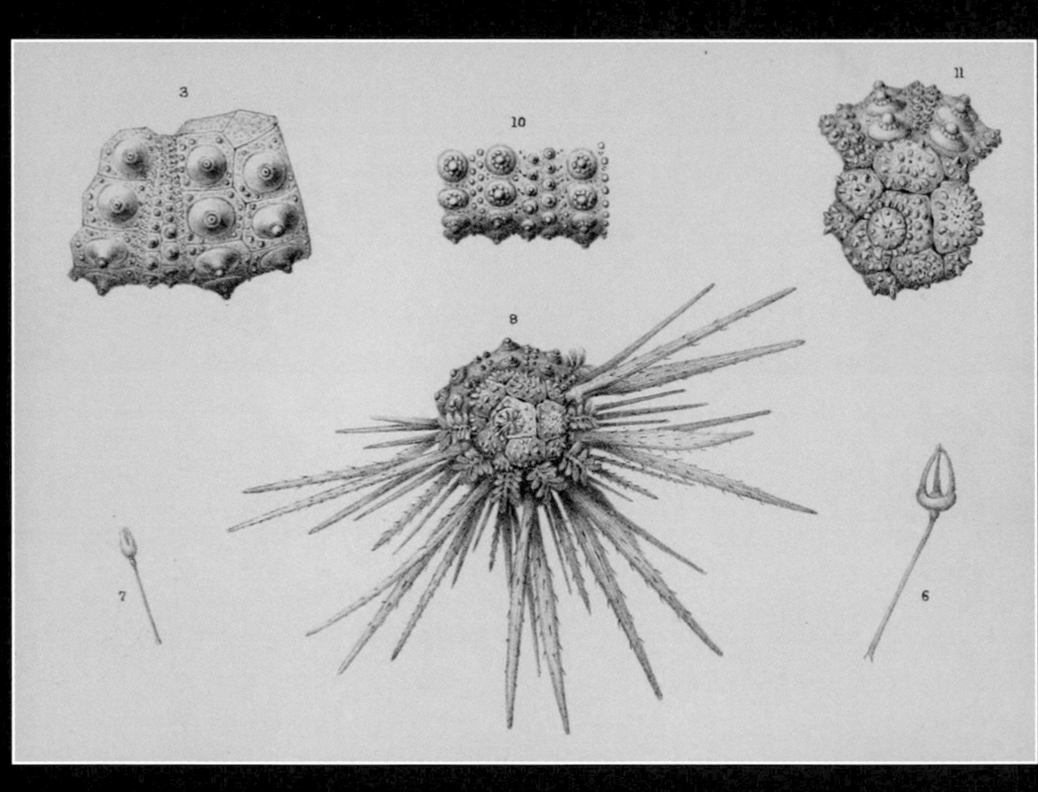
3
10
11
8
7
6

CHAPTER 6

Crossing Borders: Studying *Salenocidaris varispina*

> I felt when I got through that I never wanted to see another sea-urchin, and hoped they would gradually become extinct. Let me know of the safe arrival of the manuscript, for if anything happens to that I shall lose the little hair I have left.
>
> Letter from Alexander Agassiz to John Murray, November 1880[1]

Late on the evening of 24 May 1876, after an absence of three and a half years, *Challenger* reached Portsmouth and again anchored in English waters. The following weeks were full of excitement and activity at the dockyards in Sheerness and Chatham, as the ship's equipment and remaining supplies were unloaded. William Spry recalled on 12 June that there was a last hearty 'goodbye!' and a shaking of hands before the crew departed for home, friends and family.[2] Although the voyage at sea had finished, the scientific project carried on from a new base, a Georgian home in Edinburgh dubbed the '*Challenger* Office'. To study the mass of materials and data collected by the expedition, Charles Wyville Thomson reached out to his friends and colleagues including Alexander Agassiz, a naturalist at the Museum of Comparative Zoology at Harvard University in Cambridge, Massachusetts, who shared his passion for exploring life in the deep sea. Together, they worked to distribute *Challenger* specimens to the top specialists of the day, regardless of their nationality. Agassiz's interest in a sea urchin named *Salenocidaris varispina* (image 106) reveals a fuller picture of the initial hurdles, travels, years of collaborative research and places on land required to acquire

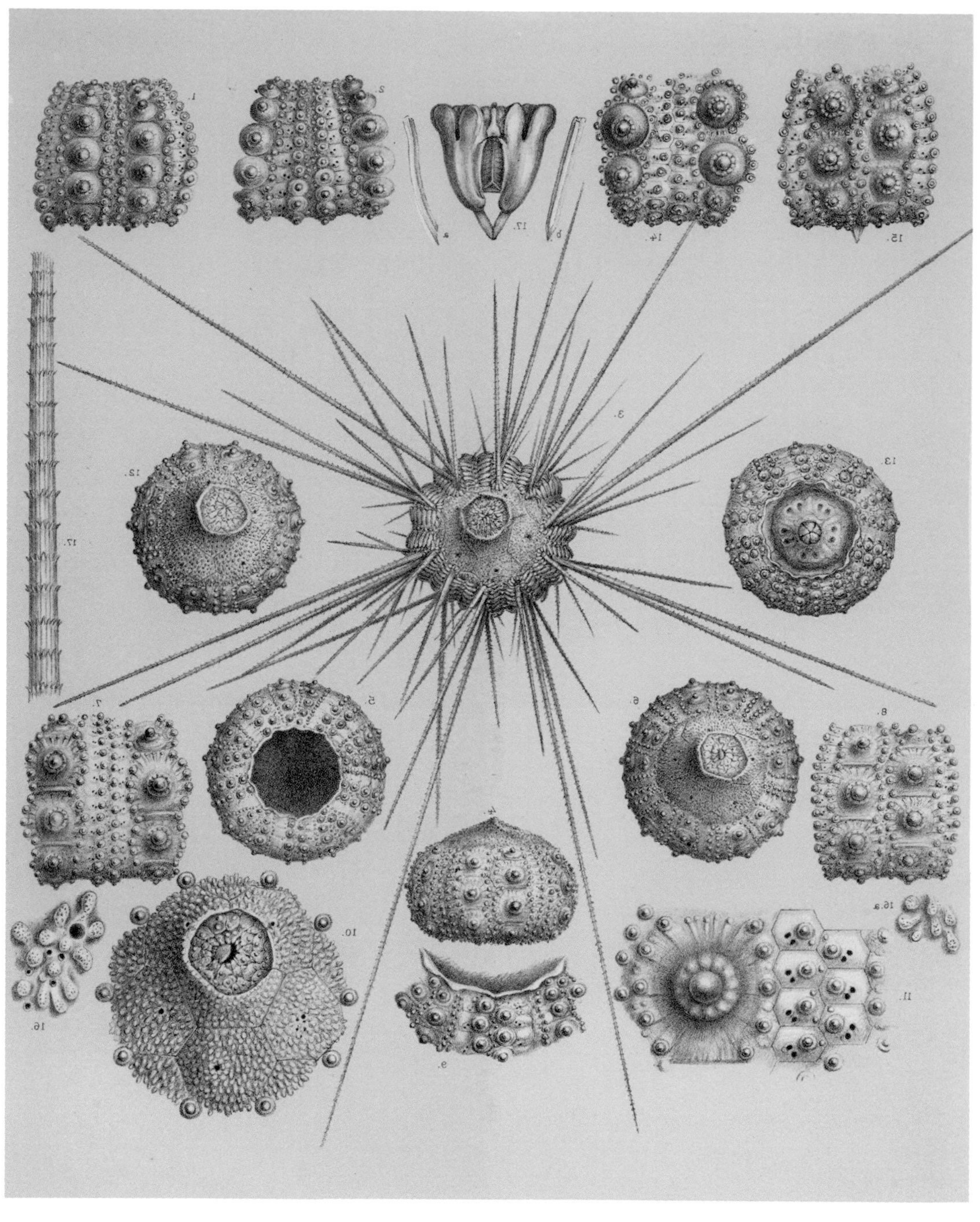

106 In the early 1870s, American naturalist Alexander Agassiz became one of the world's top experts in the study of marine invertebrates, especially sea urchins such as *Salenia varispina*, shown here in an illustration published in the *Challenger Report*.

new knowledge of ocean life.[3] Of the 50 volumes of the *Report on the Scientific Results of the Voyage of H.M.S. Challenger*, 40 were dedicated to zoology. The series, published from 1880 to 1895, connected the work of 76 scientists in Britain, Europe and America, creating the intellectual foundation for a new scientific field, deep-sea biology.[4]

Social Networks of Science: Alexander Agassiz's Trip to Europe, 1869–70

Each of the dozens of naturalists who authored *Challenger* reports had their own unique biography, but Agassiz's life offers a glimpse into some of the obstacles to undertaking marine zoology in the 1860s and '70s. Naturalists often had to rely on their own wealth to carry out scientific study and were supported by their families and wider social networks. The story of *S. varispina* thus begins with Agassiz's quest for financial independence and the means to pursue his original research.

Agassiz's father, Louis Agassiz, a professor and well-known naturalist, was director of the Museum of Comparative Zoology at Harvard.[5] Yet he had little capital, and the younger Agassiz had to financially support himself. After receiving an engineering degree, Agassiz went to work for a brief time on US Coast Survey vessels. During trips along the California coast, he created sketches and observations of marine life. In 1860 he returned to Cambridge, Massachusetts, seeking a better-paid role that would provide him a suitable income to marry the woman he loved, Anna Russell, the daughter of a wealthy Boston merchant.

Close social ties existed between Boston's elite families and the city's cultural institutions, a relationship that aided Agassiz's cause. Theodore Lyman was a philanthropist and a naturalist; he had inherited a large sum from his father and was a classmate of Agassiz's at Harvard (image 107). There was also a family connection, as Lyman was married to Anna Russell's sister, Elizabeth. Lyman made a generous donation to the Museum of Comparative Zoology in 1860 whereby Agassiz was given a salary of $1,500 a year. It was a happy time for Agassiz and Lyman wrote, 'I do not know of any married man who could look back on better years.'[6] Meanwhile, Agassiz enrolled in the Lawrence Scientific School at Harvard once more, this time studying natural history and focusing his attention

107 Alexander Agassiz (standing) with his friend and patron, naturalist Theodore Lyman (left), joined by natural history illustrator Jacques Burkhardt (right), unknown photographer, *c.*1860.

on the museum's expanding collection of marine invertebrates.

By 1865, he had reached another turning point in his career. Although he had published scientific papers and co-authored a popular book, *Seaside Studies in Natural History*, with his stepmother Elizabeth Agassiz, he found his museum salary insufficient to support his growing family and research ambitions.[7] In a scheme to increase his wealth, Agassiz invested in two mines on Michigan's Keweenaw Peninsula, an area rich in copper that projects into Lake Superior. The US government had negotiated the mineral rights to the land from the Ojibwe people in the 1840s. Despite the great quantity of copper beneath the surface, the Calumet and Hecla mines were poorly managed and facing bankruptcy. At the behest of the company's director, Agassiz went to Michigan and applied his engineering skills to redesign infrastructure such as shafts and railroads and better organise the developing mining town. After a year and a half, in 1869 the operation began to make a profit, but not without consequences. The harsh winters and frantic pace of work had damaged his mental and physical well-being, so much so that he feared he was dying.[8]

His ill health notwithstanding, Agassiz's scientific prospects remained bright. Upon returning to the museum, he resumed his intellectual life and published over a dozen articles in American scientific journals on topics from the flight of butterflies to the habits of beavers he had observed in Keweenaw. Yet he was most excited by the results of the 1868 dredging expedition of the US Coast Survey steamer *Corwin* off the coast of Florida, a voyage that opened up a new world of sea urchins, starfish, corals, crustaceans and molluscs living in a depth of 500 fathoms (914 m) and under great pressure.[9]

During this time, Agassiz gravitated towards the study of sea urchins, spiny, globular animals of the class Echinoidea, a subset of the phylum Echinodermata.[10] In the winter of 1868–69, he began work on an in-depth analysis of this varied taxonomic group. He wrote to Charles Darwin and expressed his hope that a 'careful study of such a small group as the Echini', found on the seabed of every ocean and at all depths, could provide empirical evidence for the theory of evolution.[11] Publishing such an exhaustive monograph for the museum was an ambitious and arduous undertaking that would require time, dedication and, as always, money.

For his plan to succeed, Agassiz first had to familiarise himself with all the known species of sea urchins. Naturalists did this by examining 'holotype' or 'type' specimens, the first individual of a species formally described in scientific literature. As a representative of the species, type specimens acted as a reference library for scientists to compare and classify other plants and animals, a technique that is still used today. Given their importance and rarity, the specimens that Agassiz wished to see were not loaned as 'ordinary' specimens; type specimens were carefully guarded in national museums, universities and private holdings. Agassiz was familiar with the collections in the United States, but he also needed to examine the type specimens in Europe and he lacked the necessary funds to travel overseas.[12]

A wealthy friend again came to Agassiz's aid. James Lawrence — son of the New England industrialist Abbot Lawrence who established the Lawrence Scientific School — offered to sponsor a trip to Europe 'for a rest and a change of scene', a nineteenth-century cure for many ills. Through this gift, Agassiz, his wife Anna and their two young boys travelled abroad for a year, arriving in Europe in September 1869 (images 108 and 109). The couple's scientific and social lives remained intertwined and the trip was recounted as 'a period of convalescence and of great pleasure and enjoyment; it was also a period of great activity and hard work'.[13] It was to his benefit that Agassiz carried letters of introduction to his father's scientific friends, including Darwin. In this way, he 'made the acquaintance of nearly every

108 Portrait of Anna Russell, who accompanied her husband Alexander Agassiz during his tour of European natural history museums in 1868–69. Photographer and date unknown.

109 Portrait of Alexander Agassiz, *c.*1860, who first met Charles Wyville Thomson in Belfast, sparking a friendship that shaped the international character of the *Challenger Report*.

working naturalist in Great Britain, Scandinavia, Italy, Germany, and France. Everywhere he was received as the son of an old friend'. Agassiz spent days examining sea urchins and then sightseeing with Anna; during the evenings the couple dined at one of the scientific societies or with their new acquaintances.[14] These events would later come to bear on the *Challenger Report*, as one of Agassiz's first calls was to Thomson, then a professor at Queen's University Belfast.[15] Agassiz and Thomson had been in correspondence for years concerning deep-sea life, but this was the first time they had met in person. John Murray recalled, 'Agassiz was, of course, anxious to learn all about the *Lightning* and *Porcupine* expeditions, in which Thomson had taken part'.[16] The materials from these cruises, some dredged from an unheard-of depth of 2,500 fathoms (4,572 m), were amassed in Belfast and attracted widespread scientific attention. There, Thomson, Gwyn Jeffreys and William Benjamin Carpenter allowed Agassiz to examine the echinoderms.[17]

With this access, Agassiz facilitated and accelerated the flow of information related to recent British and American deep-sea expeditions. Full descriptions of the newly discovered species would be published in scientific journals, but the process of expert analysis, illustration and printing took several years.[18] Not only did Agassiz inspect the animals collected by *Porcupine* and *Lightning*, but he also brought with him unclassified animals dredged by Louis F. de Pourtalès from the US Coast Survey steamers *Corwin* and *Bibb*. This enabled Agassiz to compare directly animals from the deep waters off the southern coast of Florida with those collected in the North Atlantic, bringing together some of the earliest deep-sea hauls acquired by British and American vessels.

On 22 October 1870, after his family's tour and his scientific studies, Agassiz left Anna and the children in London while he went to Ireland to 'see the plunder of the last *Porcupine* Expedition' in the Mediterranean.[19] After a nearly eight-week cruise, on 8 October 1870 the *Porcupine*'s captain, Edward Killwick Calver, and Carpenter had arrived at the Isle of Wight and the expedition's collection of specimens was sent to Belfast.[20] The partnership and friendship Agassiz enjoyed with the British naturalists was evident during his second visit. Speaking with Thomson about which experts would study *Porcupine*'s specimens, Agassiz recalled that they 'agreed upon the wisdom of letting the same people work up all the deep sea things from both sides of the Atlantic, as far as practicable', a decision that established a mode of working that would be adopted later by the editorial directors of the *Challenger Report*.[21] Before leaving, Agassiz wrote to Elizabeth about his success and joy at acquiring previously unstudied marine animals: 'I have picked out while here a fine series of the Echini they have collected.'[22] The British naturalists also promised to send him 'duplicates (the first series) of all that they have in way of Corals, Crustacea, and Mollusca, and Echinoderms, as fast as they are worked up and ready to be distributed', adding deep-sea gems to the Museum of Comparative Zoology's permanent collection.[23]

In the early 1870s, while *Challenger* embarked on the first part of its circumnavigation, Agassiz continued to build upon the relationships he had initiated during his scientific travels. From 1871 to 1872, Louis and Elizabeth Agassiz accompanied the *Hassler* Expedition from Boston to San Francisco around the coast of South America (image 110).

When the *Hassler* returned, the younger Agassiz sent the dredged materials to American and European experts, further strengthening a growing international community of marine zoologists interested in life in the deep sea. For his part, Thomson continued his correspondence while on board *Challenger* and invited Agassiz to visit the ship at Halifax in May 1873. There, Agassiz met the rest of the *Challenger* scientists for the first time. The naturalists were impressed by his enthusiasm and wide-ranging knowledge of echinoderms, describing him as 'cheerful, confident, and possessed a fund of dry humour'.[24]

By the end of 1873, Agassiz finished his immense work, *Revision of the Echini*. Published in four parts with 94 illustrative plates, the volume described all the known species of the class (images 111 and 112).[25] With attention to embryology and various stages of development, Agassiz omitted creatures that he found to be younger or older forms of already documented species. He also added animals that had been identified during the last two decades of ocean

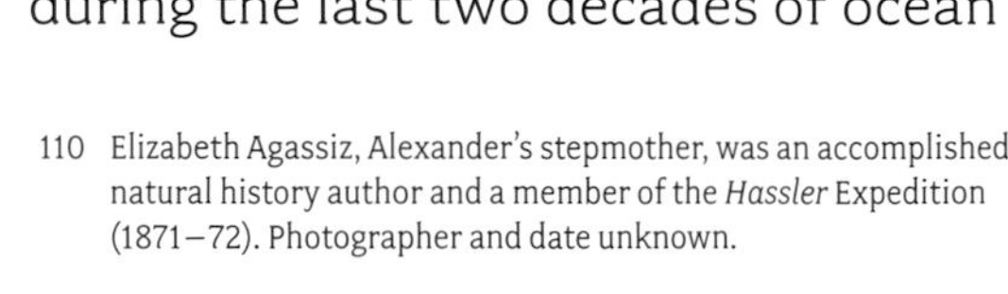

110 Elizabeth Agassiz, Alexander's stepmother, was an accomplished natural history author and a member of the *Hassler* Expedition (1871–72). Photographer and date unknown.

exploration. This all-encompassing study of the taxon assembled and unified information from multiple expeditions. The approach would be repeated in the making of the *Challenger Report*.

The volume stands testament to how knowledge of marine zoology in the late nineteenth century was generated through social networks, including Agassiz's family and friends in Boston and his subsequent connections in Europe. Although *Revision of the Echini* failed to present evidence for the process of evolution, the study raised important epistemic questions. Agassiz came to believe that what scientists knew about marine life was limited by the small number of species that had been discovered and defined so far. Overall, the monograph was exceedingly successful, propelling Agassiz's reputation as a serious marine zoologist, both in the US and Europe, and establishing him as the leading expert of the class Echinoidea.

Preparing specimens at the *Challenger* Office, 1876–77

When *Challenger* returned to Britain, Thomson confronted various problems, only some of which he had anticipated. The most pressing was the preparation of the *Challenger* collection for distribution to specialists. The expedition had sent some 600 cases to Edinburgh, containing an abundance of material — Thomson estimated there were more than 100,000 individual specimens.[26] Packaged during the voyage, animals were preserved until they could be examined in detail. Bottles of alcohol had been carefully labelled with date, depth and location, but held a motley assortment of what was brought up by the dredge. Thomson insisted that to avoid confusion, 'The specialists who undertake the different groups must see the whole of the specimens of each group.' In other words, before the naturalists could begin their studies, the entire assortment had to be separated by taxonomic group.[27]

Knowing the scale of the task, Thomson expected that the first phase of the project would take a year to complete. From the time of the ship's arrival at Sheerness, Thomson aimed to get 'the collections roughly arranged and put into a condition for distribution to experts, on the 1st of May 1877, but not before that date'.[28] In Edinburgh, he retained a close connection to the university and the city itself was a pre-eminent centre of intellectual thought. The *Challenger* Office situated at 32 Queen Street was only a short walk from the Royal

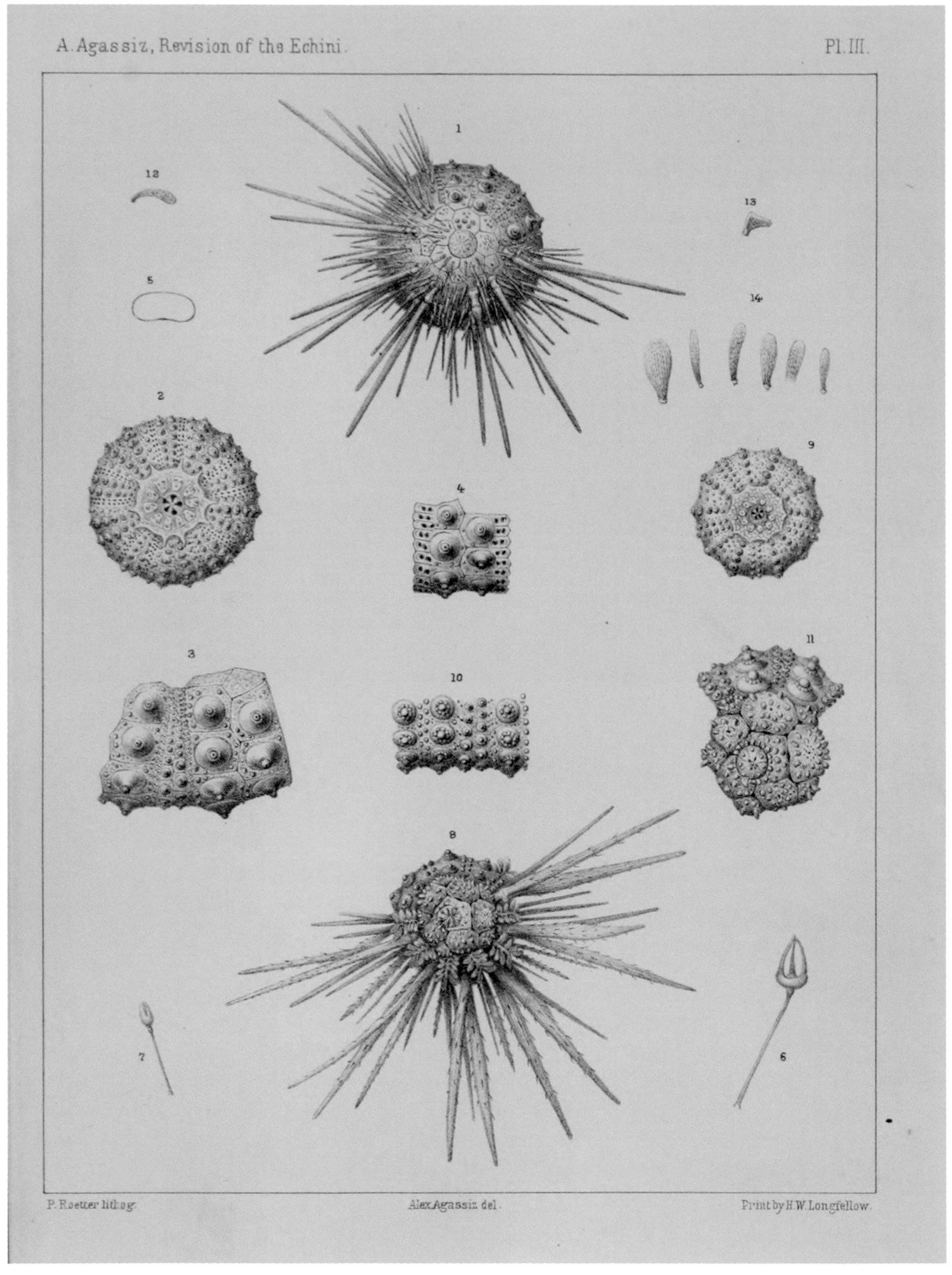

111 Alexander Agassiz reclassified and changed the name of a white sea urchin to *Salenia varispina* while compiling information for his report *Revision of the Echini.*

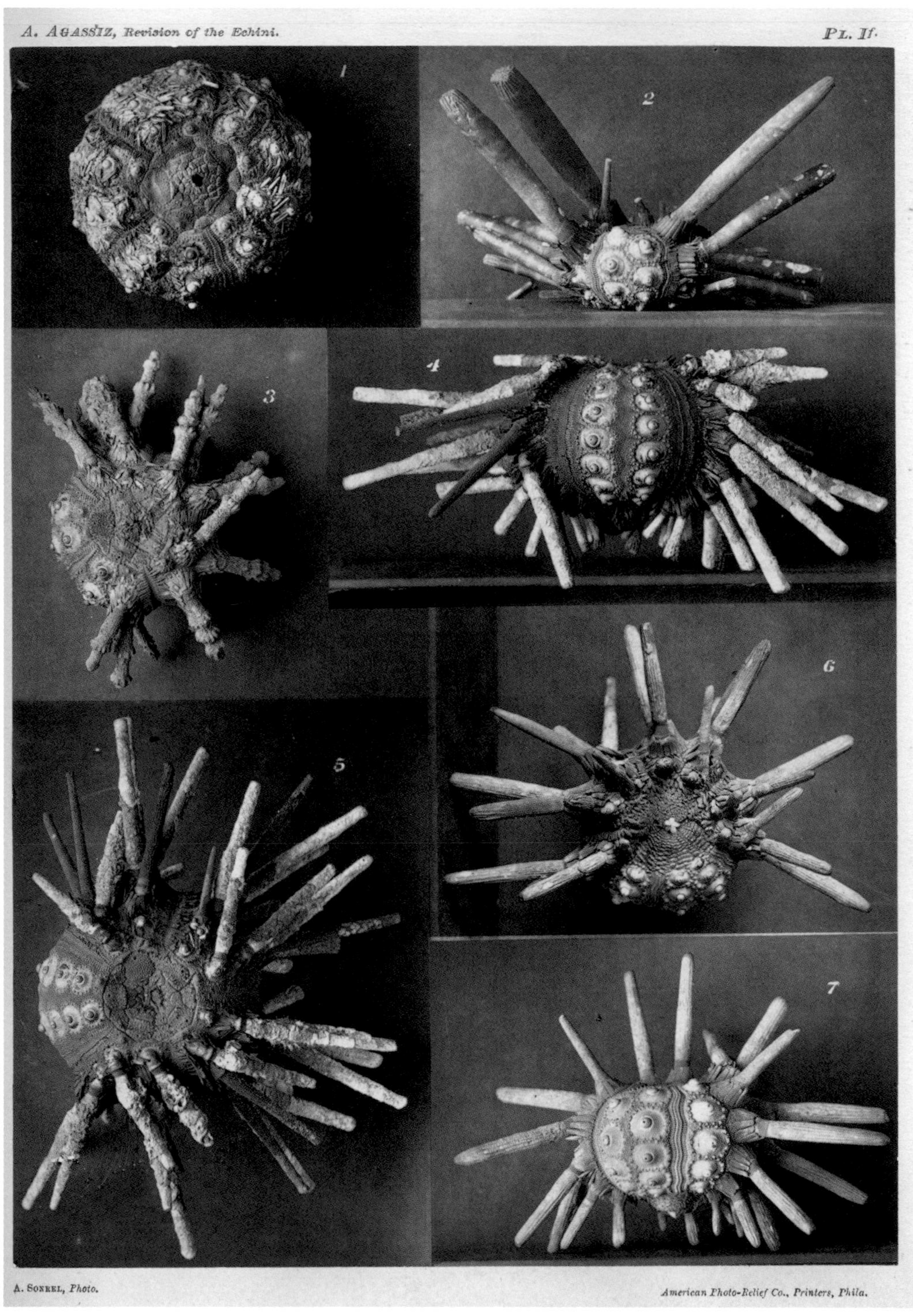

112 *Revision of the Echini* contained numerous photographic plates in addition to detailed illustrations of Echinoid specimens. The photographs shown here were captured by Antoine Sonrel and printed by the American Photo-Relief Printing Company.

Society of Edinburgh where Thomson and John Young Buchanan, both elected Fellows, presented and debated their findings. The three-storey townhouse was intended as a place where specimens could be inspected, classified, packaged and sent to specialists, and as an office from where Thomson could organise the publication of the results.

Thomson had secured a government grant to cover the rent and to engage the civilian scientific staff for an additional year. The staff salaries remained the same, with £2,000 allocated to Thomson, John Murray, Henry Moseley, Buchanan and the artist John James Wild. Frederick Pearcey continued in his role as assistant. Thomson budgeted an additional £700 for 'the pay of a servant, the putting up of fittings, and other carpenter's work, the purchase of a large quantity of glass jars and spirits of wine, travelling expenses, etc.', all necessary expenditures to convert 32 Queen Street from a large family home into a functioning naturalist's laboratory.[29]

Like the previous gathering in Belfast in 1869, scientists travelled to Edinburgh to observe and discuss what *Challenger* had discovered. Many of the leading figures in the fields of zoology, evolutionary thought and comparative anatomy became involved with the *Challenger* collections during this time, including the German zoologist Ernst Haeckel, and the English biologists Thomas Henry Huxley and Ray Lankester.[30] Evidence of life in the deep sea had been acquired through previous voyages, but the great quantity of *Challenger* materials further broadened knowledge about Earth's most expansive environment.

The *Challenger* Office brought naturalists together, but materials still had to be sent elsewhere. To help sort the collection, Thomson wrote to Agassiz and asked him to come to Edinburgh.[31] Agassiz accepted eagerly and, after a delay, arrived in late 1876. In contrast to his happy countenance in Halifax, however, he appeared despondent. While *Challenger* was on its course, he had experienced personal tragedy from which he never fully recovered. His father had died on 14 December 1873 and, eight days later, his beloved wife Anna succumbed to pneumonia.[32]

All the while, Agassiz had become a wealthy man and had realised his scientific ambitions. The Calumet and Hecla mines were now the largest producers of copper in the United States and with his new-found prosperity Agassiz could freely pursue his marine research.

Inspired by *Challenger*'s example, he had the funds to organise oceanographic expeditions on American vessels. Yet in a rare expression of emotion, Agassiz wrote to Huxley, 'How gladly I would exchange all that I have for what I have lost.'[33] He was, however, imbued with a renewed sense of purpose working with the *Challenger* scientists in Edinburgh that he had not felt since his wife's death. Agassiz threw himself into the task at hand, 'if not with pleasure at least cheerfully'.[34] Besides his dedication to the scientific undertaking, it was friendship that enabled Agassiz to carry on.

Thousands of bottles of specimens filled several rooms of the *Challenger* Office, each needing attention and identification. Of the two months Agassiz spent in Edinburgh, Murray recalled:

> Every day from early morning till as long as daylight lasted, he assisted me in opening boxes and bottles and in separating out the various groups of marine organisms, especially selecting the Echini which he was to take to America, having consented to describe this group of organisms for the Report on the Scientific Results of the Expedition. While this work was going on we had abundant opportunity for discussing the work and results of the expedition and every aspect of the new science of the sea.[35]

Once taxonomic groups were prepared for transit, Thomson sent materials to the same network of scholars that he and Agassiz had fostered since the return of the *Porcupine*. The Radiolaria (a type of single-celled zooplankton) and deep-sea Medusae (jellyfish, now classified within the subphylum Medusozoa) went to Haeckel, and the Porifera (sponges) to another German, Oscar Schmidt. Both were zoologists who had authored treatises on these animals.[36] Agassiz took with him to the United States two groups: the Echinoidea and the Ophiuroidea (brittle stars). The latter was to be investigated by Lyman, his friend and colleague at the Museum of Comparative Zoology. One of the specimens Agassiz selected was a sea urchin named *S. varispina* dredged by *Challenger* off the coast of Barra Grande, Brazil.

Thomson's decision to send *Challenger* specimens out of the country was not unopposed, and a brief but fierce debate ensued within the British scientific community. In a letter to the *Annals and Magazine of Natural History*, Peter Duncan, President of the Geological Society, argued that it was a tragedy that 'rising young

British naturalists' had been passed over in favour of those working abroad.[37] The badge of scientific prestige was at stake. Many of the animals that *Challenger* had brought to the surface were new to science. With British taxpayers supporting the cost of the expedition, Duncan argued that the honour of describing and naming these animals should be given to British scholars.

The struggle for control of *Challenger* specimens was not only about the establishment of an international scientific project, but also the growing authority of national museums as opposed to university departments and private researchers. The Natural History Department of the British Museum had greatly expanded and in 1873 construction began on an ornate new building in South Kensington. Designed as a 'cathedral to nature', the edifice had spaces for dramatic public displays and for zoologists to conduct serious investigations of the natural world. In a public letter dated 6 June 1876, Albert Günther, Keeper of Zoology at the British Museum, alleged that, 'the *Challenger* collections are not safe' in the hands of private researchers, who were more concerned with their own publications than the long-term preservation of specimens.[38] Given Agassiz's acknowledged habit of 'cutting up' echinoid specimens during his research for *Revision of the Echini*, curators perhaps had rightful grounds to be concerned for the fate of *Challenger*'s collections. Besides a worry over the safety of specimens, other major questions had to be resolved. Should scientists from other countries be allowed to claim the intellectual bounty of the *Challenger*, an expensive voyage financed by the British government? Who held more authority in this matter, Thomson or Günther?

Influential members of the Royal Society — many of whom were members of the *Challenger* organising committee — rallied to endorse Thomson's approach that respected expertise over national affiliation. On 14 June 1877 the London-based weekly journal *Nature* (founded in 1869, it remains a leading multidisciplinary science journal today) published a letter signed by Darwin, Huxley, Carpenter, Joseph Hooker and other prominent British scientists, who agreed that *Challenger*'s specimens should be distributed among those most familiar with deep-sea fauna, even if it meant they left Britain.[39]

With Parliament supportive of Thomson's position, the two factions reached a compromise. Thomson retained control over the primary objects of study, the deep-sea fauna, with the understanding that the specimens would be deposited eventually in South Kensington. Secondary materials, such as the land collections and common species of fish, were to be sent directly to the British Museum. The controversy seemingly settled, in early July 1877 Murray wrote triumphantly to Agassiz, 'We are now getting on first rate with our work.'[40] Over the following months, animals were securely packaged and shipped to marine scientists working in Britain and overseas.

Analysis at the Museum of Comparative Zoology, 1877–81

Agassiz returned to Massachusetts in early 1877, taking with him *S. varispina* and other *Challenger* specimens to the Museum of Comparative Zoology, where he was now curator (image 113). His frequent correspondence shows the robust exchange that existed between American and European researchers interested in marine life. In a letter dated 13 April 1877, Murray wrote to Agassiz:

113 The Museum of Comparative Zoology at Harvard, William Notman & Son, 1874.

> I have today sent you a small parcel containing six of our 'Station Books' and the labels from the Ophiuroids and Echini. The parcel was sent by the Globe Express. I was glad to hear that you and the specimens arrived all safe in Cambridge. I'm glad you have been able to send us the Penguins from the Pacific Coast of S. America.[41]

The *Challenger* Office distributed materials, but it also accepted many specimens in return. The penguins Murray mentioned in his letter were most likely collected by the *Hassler*. In accordance with Thomson and Agassiz's plan to better coordinate the study of deep-sea materials, the latter divided *Hassler*'s haul among 16 specialists: four were American but the remainder were Swedish, French, British or German.[42] Of these specimens, the crinoids, also known as sea lilies, were sent to Thomson for study in Edinburgh.[43] It was these continuous exchanges between voyages, people and institutions that furthered knowledge about the ocean, an environment so vast that it could not be studied by one nation alone.

In Agassiz's analysis of *Challenger* echinoids, the Museum of Comparative Zoology's biological storehouse proved an asset. After his work on *Revision of the Echini* and his travels in Europe, Agassiz wrote that the museum's collection of sea urchins represented 'with but four or five exceptions, every species described during the last forty years'.[44] Since 1860, the United States Coast Survey had sent materials to the museum for examination and, with further materials acquired through exchanges, in 1877 Agassiz claimed that second to that of the *Challenger* Expedition, the museum held 'the richest deep-sea collection in existence'.[45]

The accumulation of information about a species took time, however. For almost a decade, there was only one known example of *S. varispina*. Louis F. de Pourtalès found the first animal while dredging in the straits between Florida and Cuba on US Coast Survey steamers between 1867 and 1869. After returning from Michigan, Agassiz reviewed the collection and published a description of the sea urchin, including naming the new species. The object was then designated as a holotype and accessioned into the museum's collection.[46] During his study of *Challenger* materials from 1877 to 1881, Agassiz was able to identify seven sea urchins as belonging to the same species.[47] In this way, the museum united the efforts of past voyages with *Challenger*'s new discoveries.

The *Challenger* Zoological Reports: A Catalogue of Marine Life

Challenger's scientific reports were also enhanced by deep-sea expeditions that took place after 1876.[48] Preparing his *Report on the Echinoidea Collected by Challenger*, Agassiz included materials dredged from other ships, including the three cruises of the US Coast Survey steamer *Blake* in which he had participated from 1877 to 1880, the first year under the command of United States hydrographer and inventor Lieutenant-Commander Charles Sigsbee (image 114).[49] With his sizeable income from the mines, Agassiz was able to purchase specialist oceanographic equipment for the *Blake* to explore the ocean's depths, especially the plants and animals that lived at or near the sea floor, in the Gulf of Mexico, the Caribbean Sea and along the Atlantic coast of the United States.

After the voyage, Agassiz studied the *Blake* and *Challenger* materials together, which had the benefit of filling gaps in his knowledge from analysing one set of specimens alone. He compiled his findings in the *Report on the Echinoidea*, a volume of 321 pages

114 The deck of the *Blake*, c.1877–80. On the *Blake*, Agassiz's knowledge of machinery and approaches to hoisting heavy loads to the surface fostered new innovations. Working with Lieutenant-Commander Charles Sigsbee, Agassiz developed a system that increased the efficiency of deep-sea dredging.

and 65 plates, including an intricate illustration of *S. varispina* (see image 106, p. 170). After years of examining accumulated data and specimens, he could now determine the sea urchin was 'a characteristic species of the Caribbean abyssal fauna'.[50]

The developments in oceanography were made by a gradual amassing of information, rather than a grand breakthrough by any single voyage. Experts such as Agassiz acquired a familiarity with marine fauna through their first-hand inspection and cataloguing of animals, which were then preserved in institutional collections and published reports. As more specimens from the deep sea became available, marine scientists' understanding of taxonomic classes was revised and advanced accordingly.

Additionally, Agassiz's inquiry into the development of sea urchins, now further bolstered through his examination of the *Challenger* collection, again led him to question widely accepted views concerning evolution. In 1880, as Vice President of the American Association for the Advancement of Science, Agassiz gave an address that shocked many in the audience.[51] He had no doubt that modern sea urchins had evolved from extinct species, but declared, 'The time for genealogical trees is passed.'[52] Given the numerous defining characteristics of sea urchins — the length of spines, the complexity of teeth and the shape of the body, to name just a few — Agassiz argued scientists could not reliably map their precise lines of descent from a common ancestor, something that would only become possible with modern genome research.[53]

The drawing of elegant genealogical trees was one of the favourite tools of the 'evolutionists' and Agassiz's opinion was generally unpopular. At the core of his speech, Agassiz affirmed that there were limits to scientific knowledge. Even with his acquaintance of over 2,300 species of sea urchins, living and fossilised, he could not know all the species that had existed in the past or that were alive now.[54] His personal travel and examination of the holdings of the world's natural history museums could only take his understanding of sea urchins so far. As regards constructing a tree of life from this information, Agassiz stated, 'while we cannot but admire the boldness of and ingenuity of these speculations ... we are building in the air'.[55]

Not all contributors to the *Challenger* zoological series provided such philosophical theoretical insights, but many produced landmark studies that helped to define various taxonomic groups. Lyman completed his *Report on the Ophiuroidea* the same year as Agassiz finished his *Report on the Echinoidea*. The study was similarly robust, comprising 387 pages and 48 illustrative plates; he established 20 new genera and named 167 species. Like Agassiz, Lyman did not limit his report to animals dredged by *Challenger* and 'deemed it wise to add the names of all others previously described, and to arrange them under their genera with proper references and explanations'. By this addition, the work became an indispensable reference volume for any naturalist working with brittle stars.[56]

Despite initial protests regarding their involvement, scientists working outside Britain contributed noteworthy *Challenger* reports. Ernst Haeckel was a gifted artist and methodical scientist. His striking, geometric illustrations of plankton — microscopic organisms that live near the sea surface — were a popular sensation. Completed in 1887, Haeckel's *Report on the Radiolaria Collected by H.M.S. Challenger* was the longest report of the series. His extraordinary study spanned three volumes: two volumes of text and a third dedicated to illustrations. Within it, he identified 4,318 species of Radiolaria, of which 3,508 (nearly three-quarters of the estimated 4,700 species discovered by *Challenger* in total) were new (image 115).[57] Although he held contentious beliefs on race and eugenics, Haeckel's work as a zoologist and naturalist remains relevant to researchers today: the abundance of tiny creatures found in rocks and their long geologic history make them an important record of past marine environments.[58]

Günther's fear that British scientists would be excluded from scientific discoveries proved unfounded. Günther himself authored three reports on the fish collected by *Challenger*, including the *Report on the Deep-Sea Fishes* (1887), which extended to nearly 400 pages with 73 lithographic plates. Günther was delayed by his official duties at the British Museum, but wrote that this benefited his report as 'deep-sea explorations were being actively carried on by two institutions of the United States of America and by the Norwegian, Italian and French Governments'.[59] While these expeditions provided additional materials for analysis, they did not issue reports on the scale of the *Challenger* series. In line with other authors, Günther included in his report 'not only the species collected by the *Challenger*, but also those which from other sources are known to

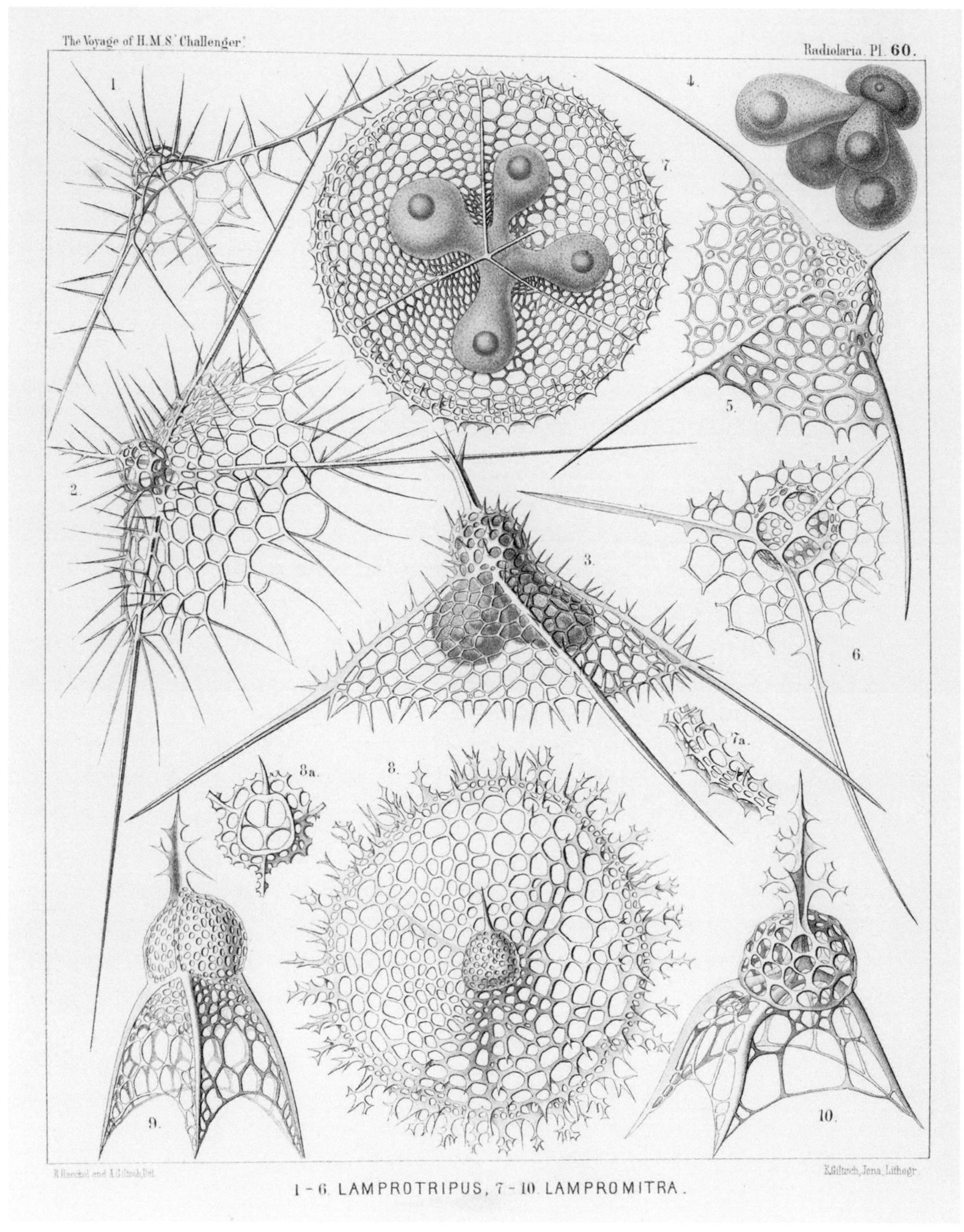

115 Illustration of *Lamprotripus and Lampromitra*, some of the 4,318 species of Radiolaria identified by Ernst Haeckel, 1887. Radiolaria are known for their intricate and beautiful skeletons, called tests, made from silica.

inhabit the deep sea'.[60] As other authors replicated this approach, the *Challenger Report* slowly manifested into an encyclopaedia of deep-sea life.

The international collaboration and a common sense of purpose allowed research to continue, even when biologists died before completing their work. During *Challenger*'s voyage, Rudolf von Willemoes-Suhm examined the Crustacea (a large, diverse taxon that includes animals such as crab, lobster and shrimp) obtained by the expedition. His work was detailed and illustrated; before his death in the Pacific, the German naturalist had already discussed and figured several striking specimens.[61] In Edinburgh, Murray sent part of the Crustacea collection to the Norwegian marine zoologist Georg Ossian Sars, one of the leading experts on this group (image 116).

Incorporating Willemoes-Suhm's notes and sketches, Sars's research focused on systematics, the study of the relationship and variation between organisms of the past and those living today. Published in 1885, his *Report on the Schizopoda Collected by H.M.S. Challenger*, which described krill and shrimp-like crustaceans, was reviewed in the journal *Science*: 'This report, by far the most important addition yet made to the Schizopoda, more than justifies

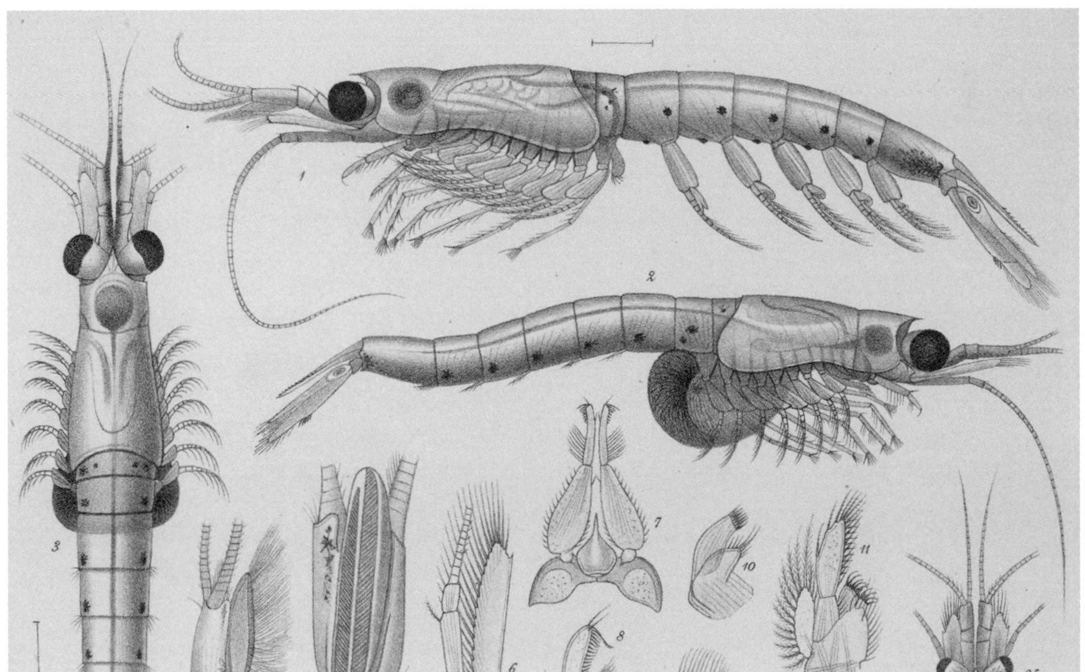

116 Illustration of the krill and shrimp-like crustaceans that Georg Ossian Sars categorised as part of the now obsolete Schizopoda order, by Rudolf von Willemoes-Suhm, published 1885.

the English authorities in intrusting [sic] certain portions of the *Challenger* collections to foreign naturalists'. Of the 57 species 'fully described and elaborately figured' in the report, 46 had been discovered by *Challenger* in widely differing regions and depths.[62]

As the reports progressed, the influence of the younger generation of naturalists such as Agassiz in the field increased. Thomson died in March 1882 at the age of 52, an event that would have signalled the end of many projects. However, through his correspondence and from his work in the *Challenger* Office since its inception, Murray had strengthened his own personal relationships with authors. When he took over the operation of the office, he reinvigorated the publishing process.

Although most of the studies considered hundreds or thousands of species, the *Report on the Specimen of the Genus Spirula Collected by H.M.S. Challenger*, published in 1895 as the 83rd and final part of the *Challenger* zoological series, focused on only one specimen, a tiny squid that lives in the deep ocean named *Spirula spirula*. The elusive animal is the only living member of its taxonomic order and an important biological link to ancient marine life forms, a subject still being explored today. It has a small internal shell resembling a ram's horn, which is used to control its buoyancy in the water. While the shells can often be found on the beach, the whole animal is seldom seen. Huxley, one of the early supporters of the *Challenger* Expedition, initially examined the animal, but due to his ill health the project was completed by Belgian zoologist Paul Pelseneer. At the time of the report's completion, only five intact specimens had ever been found. The capture of a live *Spirula* by *Challenger* was 'considered one of the happy zoological results of the voyage' (images 117 and 118).[63]

The *Challenger Report*, by now a veritable compendium of what was known of deep-sea life at the end of the nineteenth century, was a remarkable international achievement upon which future research would build. Funding remained a thorny issue for marine biologists, however, as governments were more interested for the most part in fisheries studies than bankrolling the scientific exploration of the sea. Some of the most outstanding oceanographers of the late nineteenth century privately funded specialised equipment, publications, research centres and even large-scale expeditions. The royal investigator Albert I, Prince of Monaco, devoted much of his life to the study

117 A *Spirula* (Ram's horn squid) specimen. Its internal shell, just visible, acts as a buoyancy organ and enables the animal to keep its body vertical and head facing downwards.

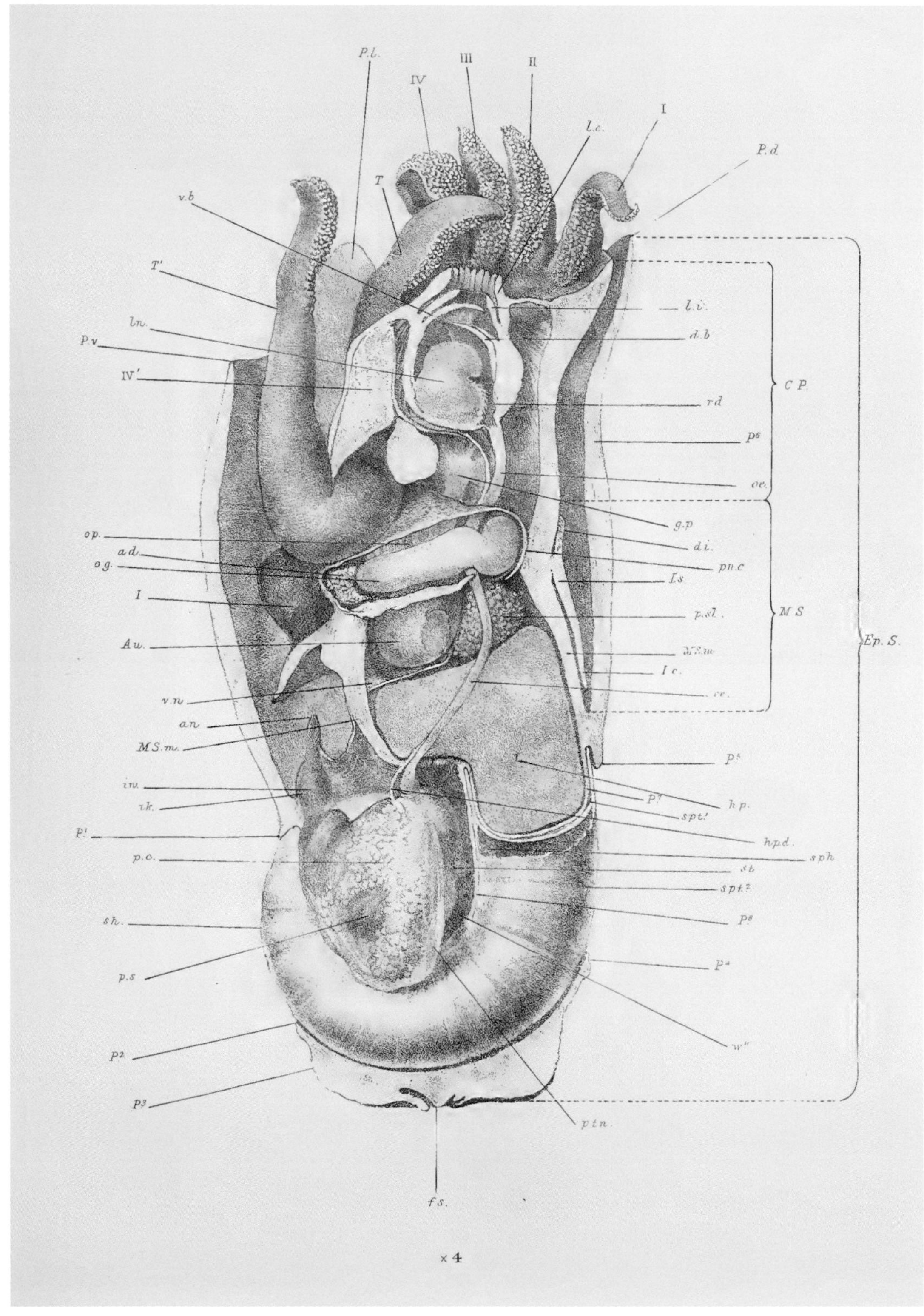

118 Anatomical drawing of the rare *Spirula* with its internal shell, by Thomas Huxley, published 1895.

of the ocean. He conducted expeditions on his impressive research yachts and in 1910 established what became the world-renowned Oceanographic Museum of Monaco, which includes a library, museum and aquarium.[64]

Like Agassiz, Murray found personal wealth through industry. Beginning in 1891 he set up a phosphate extraction facility on Christmas Island in the Indian Ocean. He had noticed the valuable deposits while examining geologic samples sent to him by the Hydrographic Office and used his contacts in government to gain permission to mine the island. By 1900, the mines were producing rich annual dividends.[65] Murray applied his substantial fortune to his oceanographic studies and in 1910 paid all the expenses for a four-month voyage across the North Atlantic on the Norwegian ship *Michael Sars*. With marine biologist Johan Hjort, he published *The Depths of the Ocean*, a book that was intended to 'present the student with a fairly complete epitome of recent advances in the modern science of oceanography', further establishing the new discipline.[66] For the rest of his life, Agassiz too continued to promote oceanographic research. He never remarried, but travelled widely and, in 1910, at the age of 75, died at sea on board RMS *Adriatic*, en route to New York from Southampton.

Postscript: Circulation Through Museum Collections, 1882–99

Upon completion of his report, Agassiz returned *S. varispina* to the *Challenger* Office. However, that was not the end of the object's journey. While the specialists involved in the publication of the expedition's scientific reports were not paid, they did receive the prestige of authorship and copies of their published work. Added to this, under certain conditions authors were granted permission to keep a selection of *Challenger* specimens for their personal use. At the Museum of Comparative Zoology, on 14 September 1881, Lyman made careful notes as he packaged up the *Challenger* specimens to be sent to Edinburgh, including *S. varispina*, and those to be retained by the museum.[67] Following Thomson's instructions, the British Museum was given a full set of the animals found by the *Challenger* Expedition, including the type specimens and any unidentified animals. Of the 'duplicates', a set was allocated to Thomson, and then, if there was enough, a set was given to the authors.[68] In a virtuous

119 The Smithsonian Institution, Washington, DC, unknown photographer, *c.*1870. In 1890, the Natural History Department of the British Museum sent *S. varispina* to the Smithsonian, where it remains today.

cycle, the *Challenger* materials therefore contributed to the authors' future studies and collections.

After being returned to the *Challenger* Office, *S. varispina* was forwarded to the Natural History Department of the British Museum. It was accessioned into the national collection in 1882, but did not stay there permanently.[69] In a letter dated 8 September 1890, Günther wrote to G. Brown Goode, Assistant Secretary of the Smithsonian Institution in Washington, DC, and suggested a group of marine invertebrates 'chiefly from the *Challenger* duplicates' for exchange (image 119).[70] Regarded as a duplicate or ordinary specimen, *S. varispina* did not hold the same value as a type specimen. However, the trade broadened the range of objects both museums had available for display and comparative research.

The museums' exchange was small, hardly a defining moment in the grand history of the expedition. Yet, it was the sum of many thousands of similar trades over the course of decades that established the great natural history collections around the world. While most of the *Challenger* type specimens are still held at the Natural History Museum in London, some were given to universities, research centres and natural history museums throughout Britain, the Commonwealth, Europe and the United States to help promote the study of marine life. For *S. varispina*, this was the end of its travels. Since 1890, it has remained at the Smithsonian where it is available for examination by scientists and curious historians of science. Over 130 years later, it is a reminder of the collaborative efforts of numerous scientists and institutions, crossing boundaries of language and nationality, necessary to explore the ocean's depths.

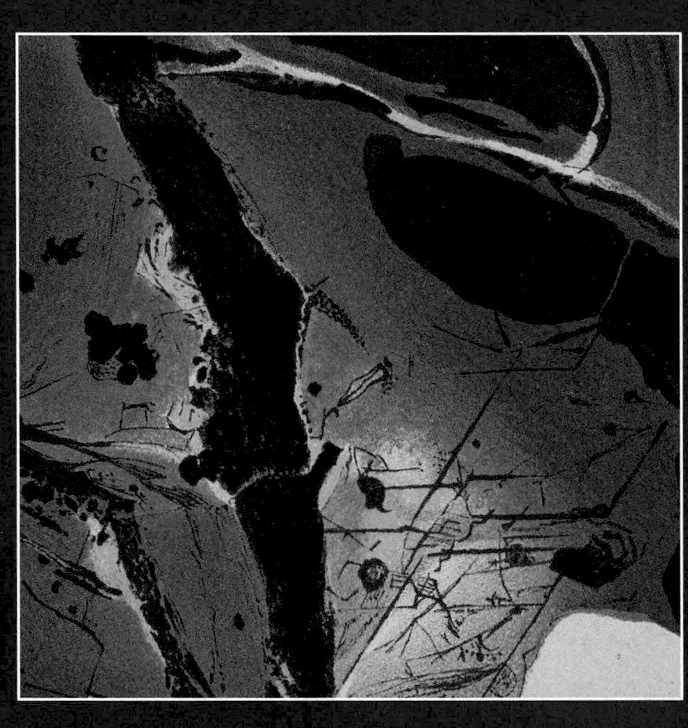

CHAPTER 7

The Art and Science of Publication: The *Report on Deep-Sea Deposits*

> We hope the completed work may be regarded as an interesting contribution to our knowledge of the ocean, and prove useful to a large number of scientific men, as it is the first attempt to deal systematically with Deep-Sea Deposits, and the Geology of the sea-bed throughout the whole extent of the ocean.
>
> John Murray[1]

When it was completed in 1895, *Report on the Scientific Results of the Voyage of H.M.S. Challenger During the Years 1873—1876* surpassed anything printed previously on the topic of the ocean. The 50 volumes of the series consisted of 30,000 pages of letterpress, 3,000 lithographic plates, 200 maps and numerous woodcut illustrations. Hundreds of people — artists, illustrators, printers, editors, naturalists, analysts, chemists and many more — contributed to the report's construction.[2] The bulk of the report describes new organisms discovered by the expedition, but the inclusion of other subjects broadened its scope and helped define the emerging field of oceanography.[3] The two volumes of the *Narrative of the Cruise of H.M.S. Challenger* (1882, 1885), compiled by Navigating Lieutenant Thomas Henry Tizard, present the expedition's hydrographic surveys and excursions on land, while also recording the methods and instruments of deep-sea research.[4] A pair of volumes devoted to 'Physics and Chemistry', published in 1884 and 1889, relate the composition, specific gravity and temperature of ocean water observed at points along the expedition's route, contributing to the study of physical oceanography.[5] The penultimate volume of the series, *Report on Deep-Sea Deposits*, authored by John Murray and Belgian geologist Alphonse-François Renard and published in 1891, contains some of the expedition's most significant scientific findings, including the first classification of marine deposits from around the globe. The making of the *Report on Deep-Sea Deposits* reveals some of the hurdles associated with committing scientific results to print, from the challenges of writing articles at sea, to the task of organising the work of teams of specialist geologists, artists and printers. In addition to textual descriptions, Murray and Renard included complex and vivid chromolithographs in their report, created through the use of microscopic petrography (image 120). Chromolithography was then an expensive printing technique that overlaid a series of different colour plates to produce a multicoloured image. Nearly two decades after *Challenger* began its circumnavigation, the volumes revolutionised the study and scientific understanding of the geology of the ocean floor.

Publication at Sea

This was a new era in scientific publishing. In 1800, printing presses had movable type but were operated by hand; the production process was slow, and books remained luxury items. Fifty years later, the steam-driven printing press transformed publishing into an industrial activity. For the first time, a book could be purchased cheaply, greatly increasing the reach of novels, newspapers and other printed material.[6] It also increased the amount and variety of printed scientific matter. Books written for an intellectually curious public became popular, while scientific journals aimed at experts were published more often and at a lower cost. Added to this, a growing network of Royal Mail postal steamers quickly circulated English-language publications between North America, Europe and throughout the British Empire, better connecting scientific researchers working in geographically disparate regions.[7] The Smithsonian Institution, for example, operated a programme of international publication exchange; in 1858 alone it sent 572 packages, weighing 9,855 pounds (4,470 kg) and containing 9,195 articles, to other organisations and libraries around the world.[8] Propelled by steam power, scientific knowledge travelled faster and farther than ever before.

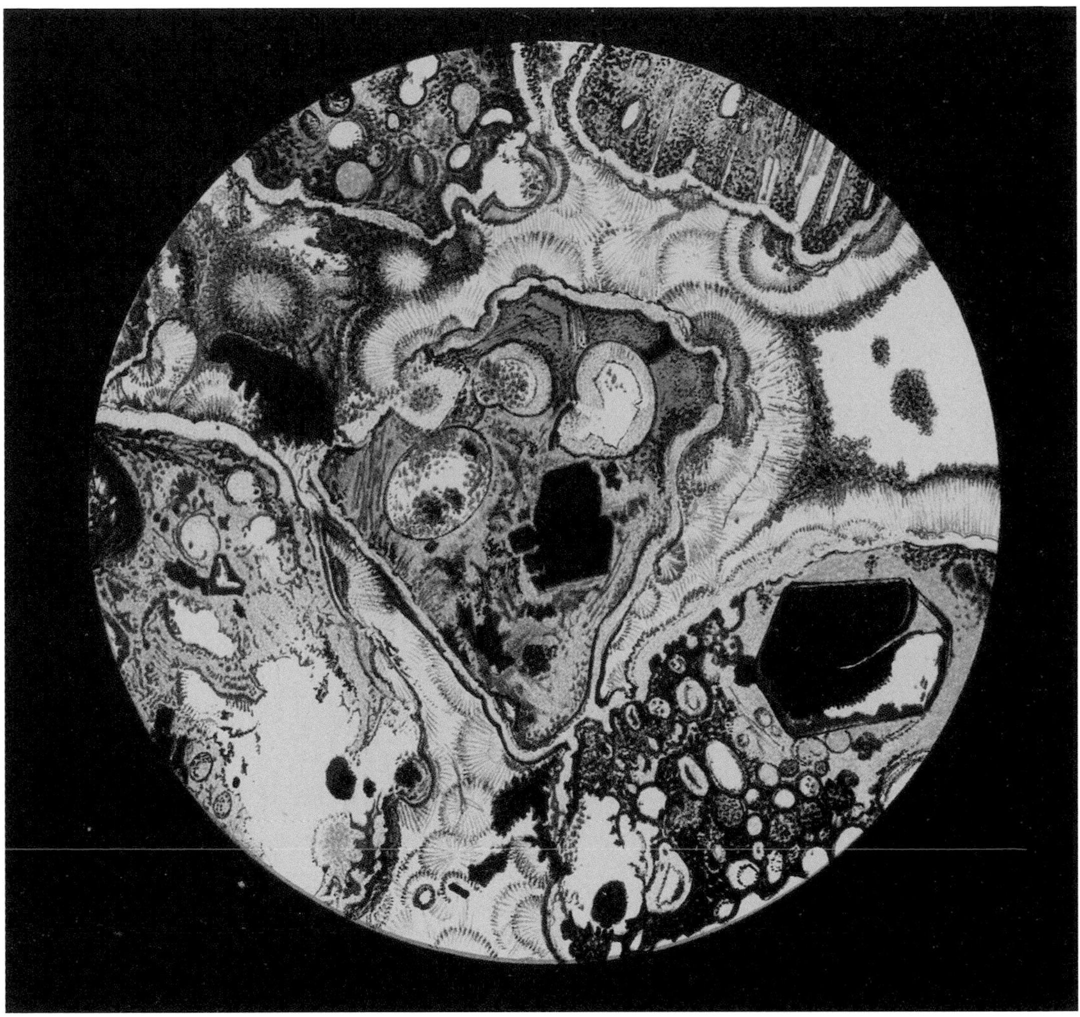

120 Section of a manganese nodule from 2,350 fathoms (4,297 m), South Pacific, reproduced in *Report on Deep-Sea Deposits*, which changed how scientists studied the geology of the ocean floor. Shown in polarised light, the specimen was magnified 145 diameters.

The seismic shift in nineteenth-century publishing influenced the work of many naturalists, including the *Challenger* scientists. As part of its ongoing support of the voyage, the Hydrographic Office forwarded journal subscriptions to meet the ship at regular ports of call. Charles Wyville Thomson thanked 'Mr. Blakeney and the other officers of the department for the accuracy with which they kept us supplied with the latest scientific periodicals, and with such instruments and books of reference as we required', thereby keeping the naturalists informed of new marine species, essential in aiding their descriptive work.[9] Besides keeping up to date with recent discoveries and philosophical debates during the voyage, Thomson penned articles describing his own work.

Whereas travelling naturalists such as Charles Darwin had previously recorded their observations in diaries and then authored a book or report afterwards (his *Voyage of the Beagle* was published in 1839, three years after the ship returned to England), it was now expected increasingly that preliminary findings would be announced as soon as possible, even while at sea. As the scientific director, Thomson felt compelled to showcase to the British public the wonders and value of deep-sea research, while updating the wider scientific community about the specifics of *Challenger*'s discoveries. In particular, Thomson wanted to publish new biological finds quickly 'to avoid as far as possible the multiplication of synonyms by the description of the same species simultaneously by different observers', a problem that was encountered on an ever more frequent basis as previously unknown life forms were rapidly being brought to light by the deep-sea sampling conducted by *Challenger* and other expeditions.[10]

Some naturalists found it easier to draft articles for publication during the expedition than others. Henry Moseley was a prolific writer who appreciated his time at sea:

> There are many worries and distractions, such as letters and newspapers, which are escaped in life on board ship, and the constant leisure available for work and reading is extremely enjoyable. I felt almost sorry to leave, at Spithead, my small cabin, which measured only six feet by six, and return to the more complicated relations of 'shore-going' life.[11]

Throughout the voyage, Moseley wrote papers on botanical subjects and described new species of marine invertebrates in a variety of publications, including two articles in Britain's oldest scientific journal, *Philosophical Transactions*.[12] Reaching a wider audience, he published regularly in *Nature*, including five articles in 1874 alone.[13]

Murray, a naturalist with expertise in geology and charged with overseeing the collection of deep-sea deposits, found the experience of preparing reports under these circumstances more arduous. Joining *Challenger* at 31 years of age, he had been a bright student at the University of Edinburgh but struggled with academic exams and did not graduate (image 121). As ship's surgeon in the whaling ship *Jan Mayen* in the Arctic Ocean in 1868, Murray collected marine specimens

121 A portrait of John Murray from 1895, the year the *Challenger Report* was completed. Unknown photographer.

but without the systematic approach required of a scientific expedition. Although he developed some of his early theories about marine deposits during *Challenger*'s cruise, he did not have Moseley's or Thomson's writing experience. Furthermore, geological samples proved hard to describe. After *Challenger*'s dredge or sounding tube had recovered materials from the sea floor, the sediments were preserved, briefly examined and then sent to Edinburgh for safekeeping.[14] In the preface to his first journal article on the subject, Murray implied that he had written it only under duress:

> The Preliminary Reports have been prepared at the request of Professor Wyville Thomson. They have been compiled during the past three weeks from notes taken daily during the past three years. In only a few cases has it been possible to refer to the objects remarked upon, they having been sent home for greater safety, or packed away beyond reach.[15]

Nevertheless, with Thomson's encouragement Murray composed his article 'Preliminary Report on Oceanic Deposits' during the long voyage from Tahiti in the central Pacific Ocean to the west coast of South America. He posted it to the Royal Society in London from Valparaiso, Chile, in early December 1875. With the advantage of transport by postal steamers, the society received Murray's report on 14 February 1876 and it was read and presented at a meeting on 16 March, two months before *Challenger* returned to Britain.[16]

Murray, Moseley and John Young Buchanan each authored an article for a special edition of the journal *Proceedings of the Royal Society of London*, while Thomson and Rudolf von Willemoes-Suhm contributed two papers each. The journal volume was illustrated by 33 figures, including a chart showing the ocean deposits observed along *Challenger*'s route (image 122).[17] Murray's article was also complemented by four chromolithographic plates. The chart, like other geologic illustrations of the time, used assorted colours to distinguish between various sediments. Murray's research, which would form the basis of his *Challenger* report, was thus in print three months before the expedition returned to Britain.

Mountains Under the Microscope

Since the early planning stages of the voyage, Thomson had advocated that *Challenger*'s results be assembled in a 'fixed form', meaning that the findings should be published and bound in an

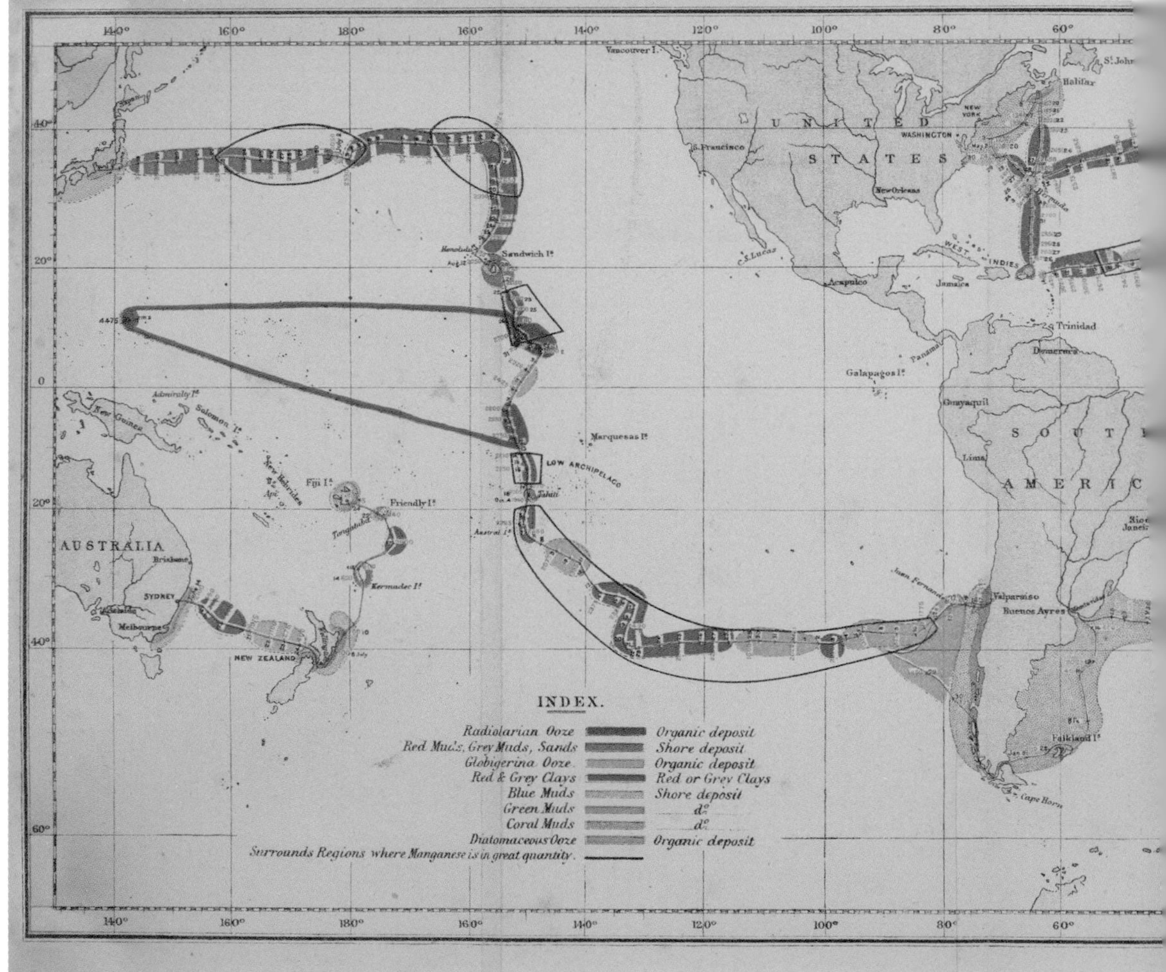

122 Chart of deep-sea sediments along the route of HMS *Challenger*. John Murray prepared his preliminary report in the South Pacific, but it was limited by the specimens and capabilities of the equipment on board.

expedition report, a series of multi-authored volumes. This was a literary genre established in the early nineteenth century and not an unusual request.[18] However, ambitious scientific publishing projects often failed, especially those that depended on public funds or a naval captain to complete. Thomson warned of 'the unsatisfactory history of similar expeditions' in which records made during the voyage became fragmented and lost over time, rendering months or years of observations almost meaningless.[19] Thomson wrote: 'It was, therefore, from the first a matter of grave consideration with me by what machinery it might be possible to keep the collection together and to prepare a report, which should justify the expense of the voyage.'[20] Even once research was completed, a lack of funding could

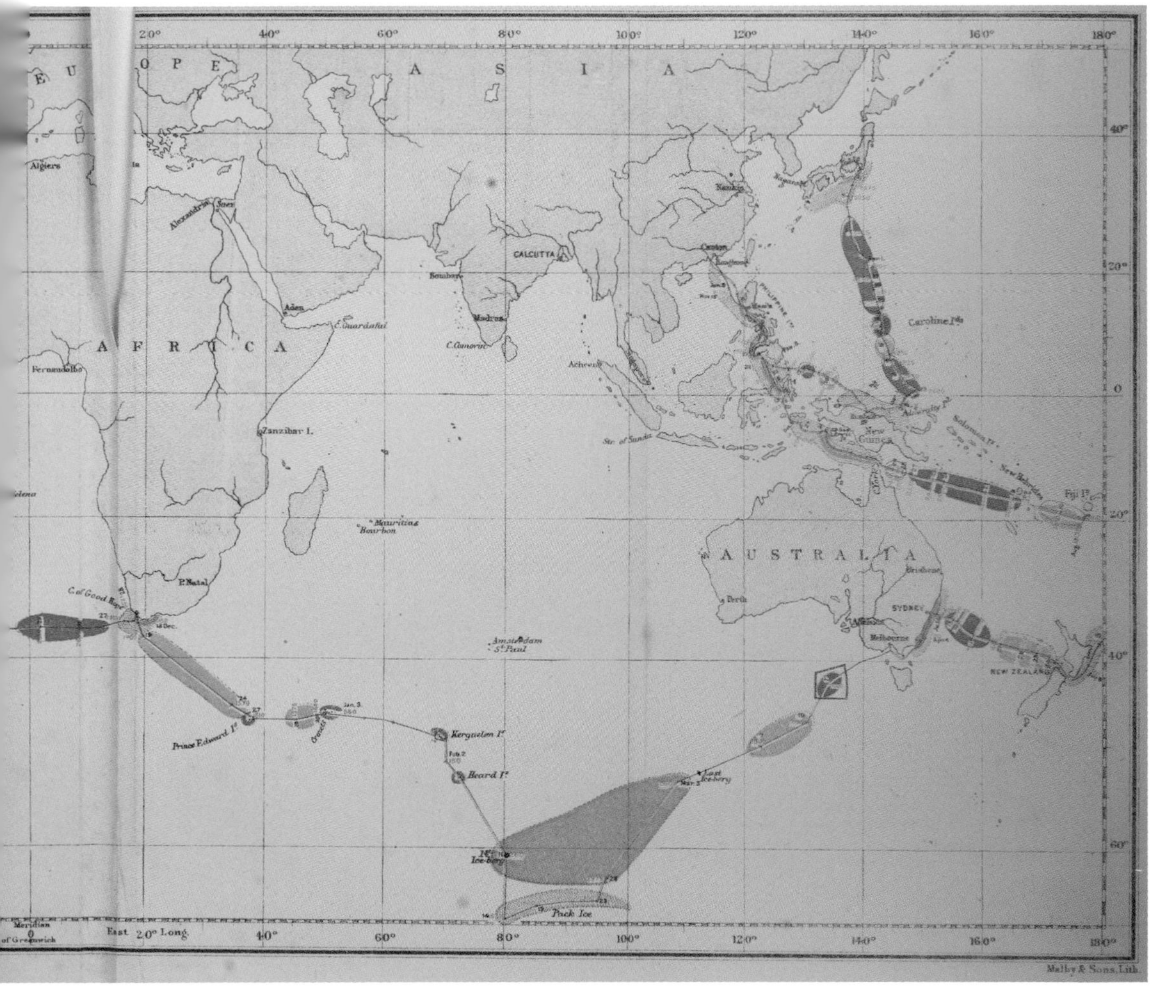

derail publishing efforts; for instance, in pursuit of publications of the highest quality, the US government decided to finance the printing of only 100 copies of the 15 scientific reports of the United States Exploring Expedition (1838–42).[21]

The 'machinery' of publication that Thomson had to organise was the most complex aspect of the expedition's scientific work thus far. During the circumnavigation, Thomson had submitted a course of action to the Lords Commissioners of the Admiralty, including details of expenditure for what he calculated was a five-year project to prepare the scientific results.[22] The government approved his general proposal and in 1876 a grant of nearly £25,000 was agreed for publication through the Treasury and Stationery Office.[23] This hefty sum enabled Thomson to begin setting into motion his ambitious plan.

After the expedition's natural history collection was sorted and sent to experts to study, in May 1877 the character of the *Challenger* Office shifted from that of a small natural history museum to a busy book publisher. Thomson's position changed to a more administrative role during this last phase of work: 'It will be obvious that my function is simply that of Editor ... and all I can undertake at the Office is to bring the series as nearly as possible into uniformity as they pass through the press.'[24]

Although most of the original scientific staff from the voyage had left the *Challenger* Office, those associated with publication now came to the fore. Thomson hired a new assistant to 'look after the correspondence and accounts, and to take on the mechanical and detail part of the editing'.[25] Murray's role also shifted to encompass more editorial tasks, such as answering correspondence from authors.

Artwork and illustration were a key part of Thomson's vision for the *Challenger* series. He was buoyed by his most recent project, a popular book completed shortly before the expedition left Sheerness, *The Depths of the Sea*, which chronicled the voyages and discoveries of the *Porcupine* and *Lightning*. As advertised on the front cover, *The Depths of the Sea* included 90 images and eight charts of various sizes.[26] Similarly, Thomson proposed that *Challenger*'s report should feature 'many lithographed plates, and many charts, woodcuts and photographs' to describe species and explain methods of working, a feature that would appeal to a wide Victorian audience, including scientific researchers and an interested general public.[27] Following Thomson's design, the *Report on Deep-Sea Deposits*, as with the other volumes, would include vivid illustrations to showcase and complement the reams of data and textual descriptions.

As the various parts of the *Challenger* series progressed, Murray continued his study of deep-sea geology. The lively intellectual atmosphere in Edinburgh proved an inspiration and he was introduced to sophisticated types of analysis by his former university professor, Archibald Geikie, one of the leading geologists in Scotland. To study and classify a rock's composition, petrographers ground down samples into thin, transparent sections and placed them under a microscope. Illuminated by polarised light (akin to the effect of polarised sunglasses), the directed light waves enhanced the contrast and quality of the image. Mineral crystals could then be readily identified by their characteristic properties such as colour and structure. Under these conditions, for example, garnet appears pink. The method was first proposed in the 1840s, but its use was limited by the quality of optical lenses available.

During the 1870s, however, new discoveries relating to the laws of optics rapidly transformed the study of geology. While *Challenger* was on its circumnavigation, a breakthrough occurred when the German physicist Ernst Abbe developed a mathematical formula (now called the Abbe sine condition) that produced a high-quality lens free from distortion. Applying this improvement to petrographic microscopes, a rock's crystalline structure was made visible as never before; scientists could now determine a rock's composition and the process by which it was formed. In 1877 a writer for *Geological Magazine* observed that 'not long ago' geologists carried in the field a handheld magnifying glass to aid in a rock's identification. Now, the character of mountain ranges was 'studied with the microscope'.[28] Earth's geologic history can be thought of as recorded in stone and Abbe's discovery enabled scientists to tell this story better.

While examining *Challenger*'s mineralogical materials, Murray expended considerable effort studying manganese nodules, a particular type of rock found in certain parts of the seabed. Generally between 1¼ and 4 inches (3–10 cm) in diameter, they grow slowly over thousands of years with concentric layers of iron, manganese and other minerals and metals forming around a core. When crushing the nodules to examine their constituent elements, he identified small magnetic particles but could not determine their origin. Viewed under a microscope, he realised the objects closely resembled cosmic spherules, particles of iron that fall to Earth from space. With Geikie's guidance, Murray was able to confirm that the magnetic iron enclosed in manganese nodules and other sediments on the ocean floor was 'cosmic dust', microscopic pieces of meteorites that survive travelling through the atmosphere (image 123).[29] This major breakthrough explained how iron, an essential element to living organisms, is found both in the deep sea and a great distance away from the influence of coastal deposits. In his paper, presented in the *Proceedings of the Royal Society of Edinburgh* in 1877, Murray thanked Geikie and his colleagues for their 'many hints' and contributions, and gave credit to Frederick Pearcey, who undertook 'much of the mechanical work which an examination of these deposits has entailed'.[30] Today, Murray's discovery of cosmic spherules in deep-sea deposits is recognised as one of the expedition's most noteworthy scientific findings.

Although microscopic petrography offered exciting possibilities for Murray's study, few British scientists were familiar with the technique at the time.[31] One barrier was that the first major works on the subject that defined the nomenclature had been written in German. While some British scientists could speak and read the language, it would take a German petrographic expert to translate correctly the new technical terms and descriptions of rock specimens into English. This breakthrough happened when Ferdinand Zirkel authored an expedition report for the US Army, written in English. From 1867 to 1872, Clarence King led the US Geological Exploration of the Fortieth Parallel that mapped 61,776 square miles (160,000 square km) across Colorado, Nevada, Utah and Wyoming. After collecting thousands of specimens in the field, King asked Zirkel to study the samples and illustrate his

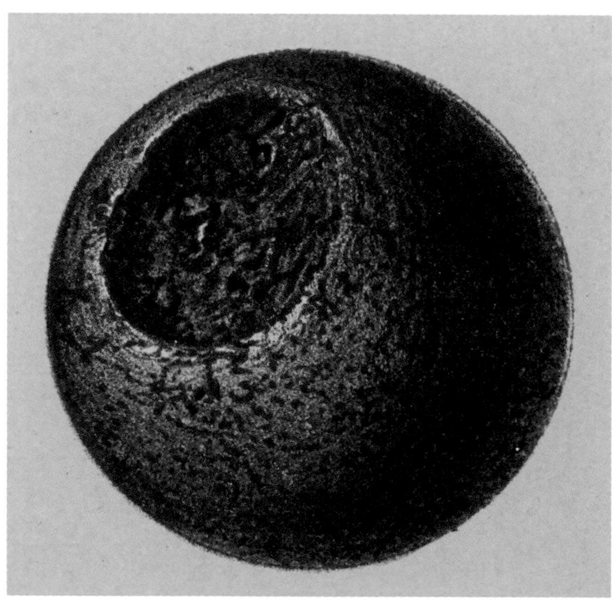

123 A chromolithographic illustration showing a cosmic spherule, one of the most significant discoveries of the *Challenger* voyage.

124 A portrait of Alphonse-François Renard, unknown photographer, 1896.

findings. Published in 1876, the finished volume, *Microscopical Petrography*, laid out a system of classification for American and British geologists, a missing piece for Murray's identification of ocean sediments.[32] The publication heavily influenced the design of subsequent geologic studies, including *Challenger*'s scientific report.

Colour was an essential factor in the identification of rocks, as demonstrated by Zirkel's descriptions and rich illustrations (image 125). To represent the mineral compositions, the figures utilised a range of hues and appear as intricate mosaics. Zirkel defined the appearance of specific rock types, from the 'black, entirely opaque, amorphous grains and scales' of opacite, to the 'yellowish, reddish, or brownish amorphous earthy substances' of ferrite.[33]

Funded by the US government, *Microscopical Petrography* was made using chromolithography and transformed the field of geology. However, this type of study was well beyond Murray's expertise. It was therefore a momentous day in *Challenger*'s history when Alphonse-François Renard, one of Zirkel's students, visited the *Challenger* Office in the autumn of 1877 (image 124).[34] A year earlier, Renard had co-authored a study of Belgian plutonic rocks using the new methods; with the costs underwritten by the Royal Academy of Belgium, his report included chromolithographs in the same vivid style as those in *Microscopical Petrography*, demonstrating Renard's talents as both analyst and illustrator (image 126).[35]

After presenting a paper in London, Renard met Thomson and Murray in Edinburgh and viewed the *Challenger* collection of deep-sea sediments. He left intrigued by the possibility of applying microscopic petrography to facilitate a better understanding of ocean processes and geologic formations, a methodology that had not yet

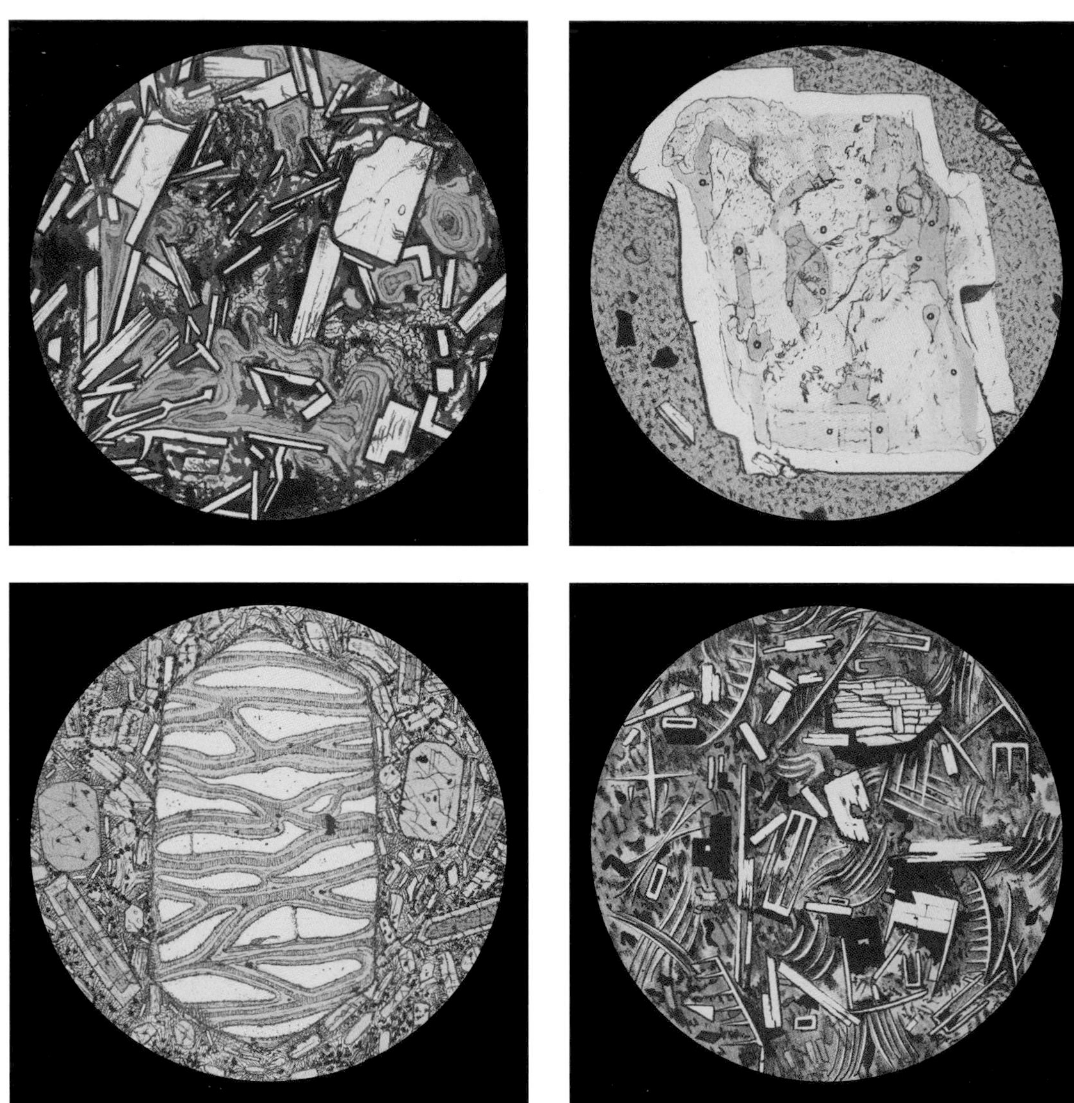

125 Ferdinand Zirkel's *Microscopical Petrography* changed how the composition of rocks was studied and illustrated when it was published in 1876, the year *Challenger* returned to Britain.

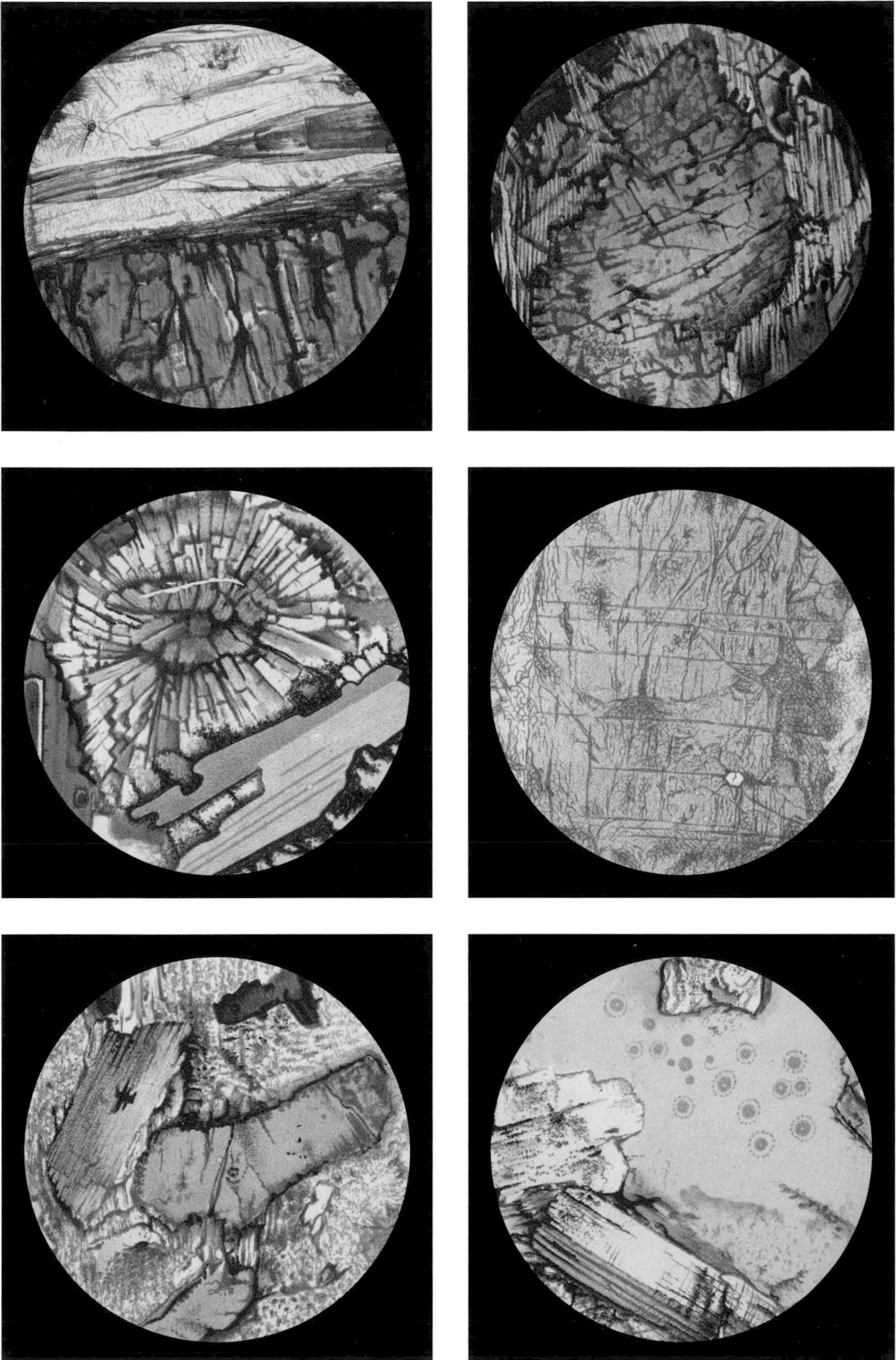

126 Illuminated rock specimens as seen under a microscope. Alphonse-François Renard's use of a solid black background brought attention to the dazzling array of bright colours found within such materials.

been attempted.[36] Thomson must have been impressed by Renard's knowledge and skills. In a letter dated 5 September 1878, he extended a formal invitation to the Belgian geologist to begin work on the *Challenger* materials, writing that he believed it would be 'important to science' for him to examine the samples and 'to contribute an account of them with suitable illustrations'.[37]

The Many Layers of Analysis and Illustration

In 1881 Murray's study of deep-sea sediments took a new direction with the addition of co-author Renard. The work of analysis, however, was an exacting and decidedly slow process that involved several stages of physical and chemical separation, examination and measurement. Preparing thin sections of rock also required considerable time, skill and specialised laboratory equipment.[38] For the first two years, Murray and Renard conducted research together at the *Challenger* Office for prolonged periods.[39] Progress was gradual and on 1 March 1881 Murray updated his friend Alexander Agassiz: 'Renard is ill here so our muds do not get on as fast as I would like.'[40] When Renard left Edinburgh to return to his laboratory in Belgium in 1882, a large number of the deposits had been examined in detail, but an enormous amount remained unseen.

To distribute the work of the study, Renard and Murray enlisted the help of other experts. Thanks to the friendships and scientific connections that he had established while undertaking his research in Vienna, Renard sent *Challenger* deep-sea materials to three specialists working in a laboratory there.[41] Renard also oversaw analysts based at his own laboratory in Brussels. Supplementing Murray's efforts at the *Challenger* Office, John Gibson conducted a geochemical analysis of the composition of manganese nodules at the University of Edinburgh. As the project gained momentum, teams of petrographers operating in three countries began to make headway.

Scientific investigation and illustration happened simultaneously, and chromolithographs were produced over time as the plates were prepared. Although the report would not be fully assembled for another eight years, Renard began reviewing images for publication as early as April 1883.[42] To create portraits of rock samples that were useful to scientists, certain objects within the optical field were ignored and others emphasised.[43] Each of the three teams of analysts therefore worked closely with lithographers, skilled artisans who transferred drawings onto stones for printing, in their respective cities.[44] Small deviations in colour were essential to mineral classification, so the selection of ink colours was necessarily precise.[45] Lithographers drew onto a flat stone (called a plate) using oil, fat or wax. The stone was then treated with a mixture of acid and gum arabic, which raised the drawing from the surface. One plate was used to print a single colour, so each additional colour increased the time and effort involved.

The etched stones were then sent to specialist printers. In Vienna, a striking illustration of cosmic spherules was printed at the Kaiserliche und königliche Hof- und Staatsdruckerei, the Imperial and State Printing House of Austria, one of the most sophisticated printing houses in Europe (image 127).[46] The same process that was used to make government stamps, involving elaborate overprinting to prevent counterfeiting, was used to create figures for Murray and Renard's *Report on Deep-Sea Deposits*. The application of 12 or more plates was not unusual in the production of a work of this calibre. Each sheet of paper therefore passed through the printing press as many times as there were colours in the final print. Depending on the number of colours present, a chromolithograph could take even very skilled workers months to produce. To avoid blurriness and overlap of the colours, each of the plates had to be aligned to precise reference points placed on the paper. This exactness can be seen in the golden yellow background of the plate printed in Vienna, which does not interfere with the detail at the edges of the figures. The illustrations of the cosmic spherules and their microstructures also contain subtle tones and colour variations. Each of the 29 chromolithographic plates that appeared in the report was the outcome of a similarly meticulous manufacturing method, using the finest-quality printing methods available.

The report's authors employed a variety of visual techniques to display their findings. Renard's presentation of manganese nodules resembled the style of *Microscopical Petrography* and were set against black (image 128). Other samples were set against lighter colours and depicted rocks drawn as three-dimensional objects. In the description of mud samples with Foraminifera, single-celled organisms vital to the marine food chain, the authors added a printed

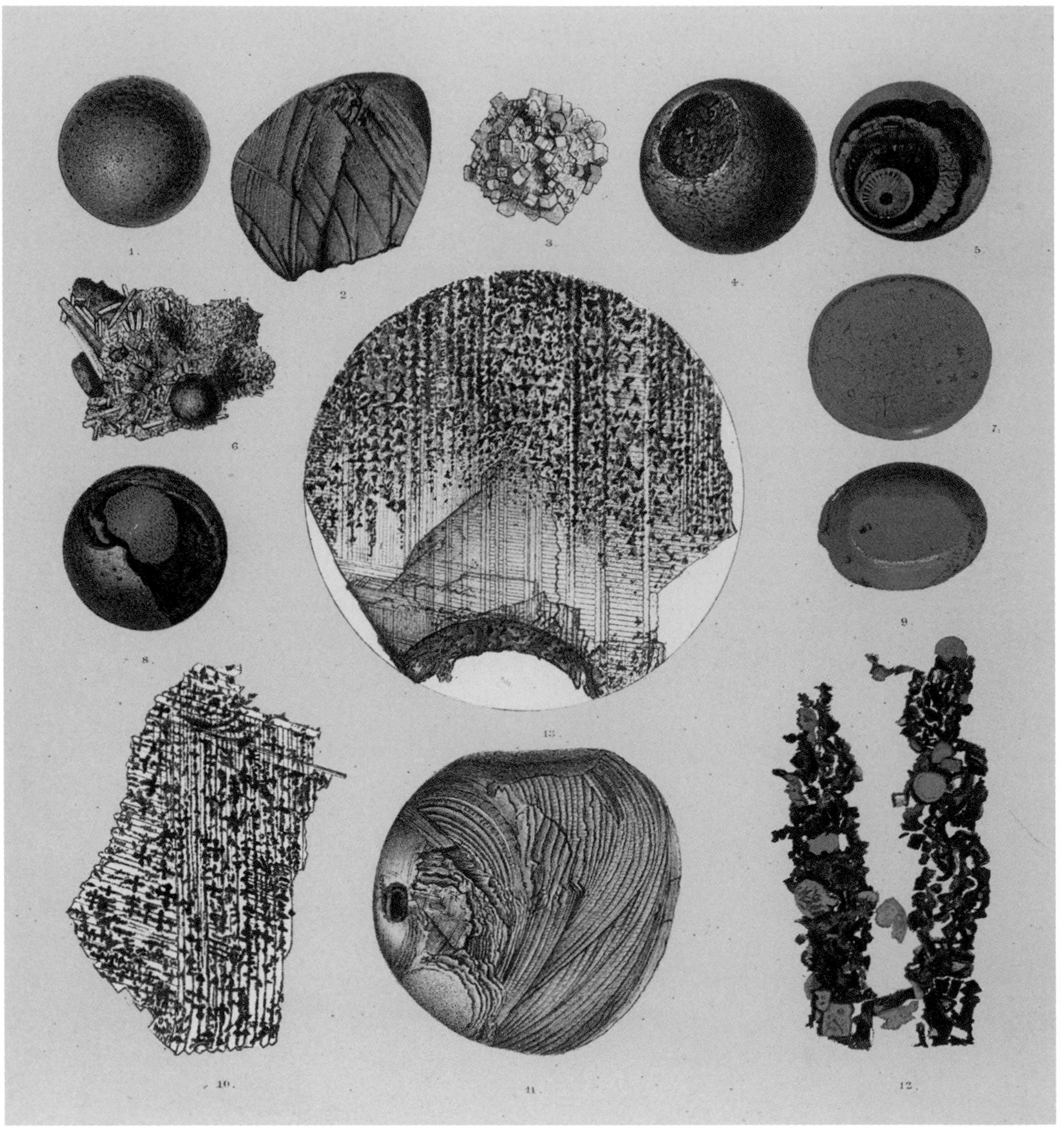

127 An assortment of cosmic spherules in various forms reproduced in the *Report on Deep-Sea Deposits*, printed at the Austrian Imperial and State Printing House, Vienna. The top left figure (fig. 1) was extracted from a manganese nodule found at Station 285, at a depth of 2,375 fathoms (4,343 m) in the South Pacific. Magnified 90 diameters the spherule has a coating of black magnetic iron with a reflective surface. The central figure (fig. 13), magnified 390 diameters, shows the microstructure of the sample below it, taken from mud found at Station 338, at a depth of 1,990 fathoms (3,639 m) in the South Atlantic.

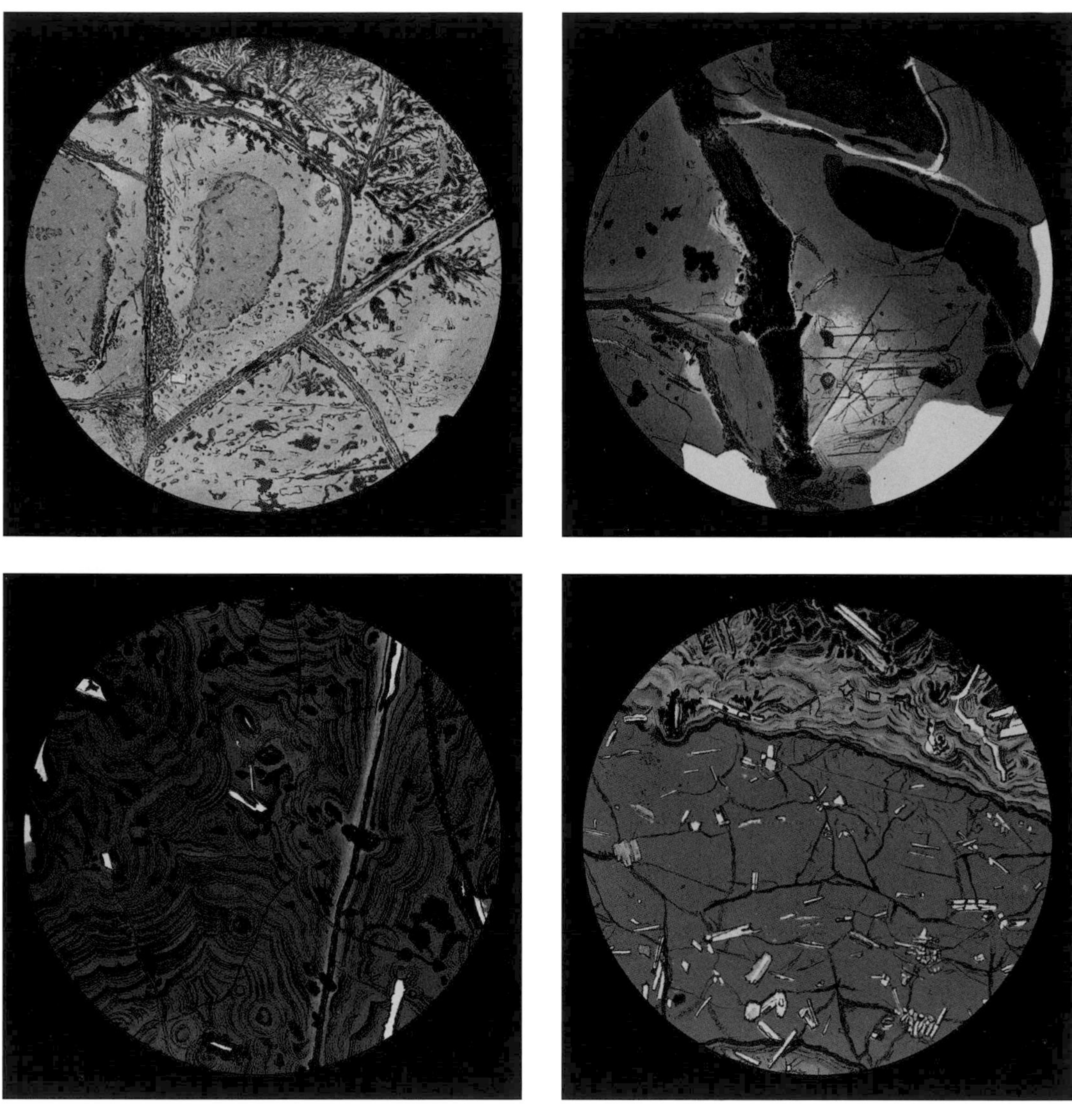

128 A mix of colours and intricate structures demonstrates the mineral composition of rocks collected by *Challenger* below 1,400 fathoms (2,560 m). The bottom left image (fig. 13) is from Station 160, from a depth of 2,600 fathoms (4,754 m) in the southern Indian Ocean. Within the nucleus of a manganese nodule was found basaltic volcanic glass and crystals of olivine and augite.

tissue overlay bound into the volume, which outlined and labelled various species (image 129). The mud obtained from the dredge would have been hardened before being cut into thin sections so that it could be viewed by transmitted light microscopy.

The Failure and Fixing of the *Challenger* Office

As work on the *Report on Deep-Sea Deposits* continued, the direction of the *Challenger* Office under Thomson became increasingly chaotic. At the end of 1881, several parts of the series were in various stages of development, but only three volumes (consisting of ten reports, including a description of green turtles and another featuring the bones of dolphins dredged from the ocean bed) had been realised. Thomson had great difficulty managing the complications posed by the large folio format, the many authors involved and the manufacture of the illustrative plates.[47] As the number of volumes in production soared, the costs of printing spiralled beyond the project's initial budget. Compounding the problem, the Treasury was hesitant to renew its funding. It was a turning point that almost led to the failure of the entire enterprise. Similar difficulties had curtailed the publication of reports of other notable expeditions, including that of the United States Exploring Expedition. As time and resources dwindled, it was becoming increasingly likely that the *Challenger* Expedition report would consist of a few scattered volumes.

After a period of ill health, Thomson died on 10 March 1882 from the 'effects of paralysis' at the age of 52. He was a popular figure, and his death came while attempting to renew a Treasury grant, a period of tremendous stress. His circle of scientific and political friends pressed for the publication of the *Challenger* results to proceed and for the renewal of government funding.[48] Although the bulk of the series was published after his death, each volume refers to Thomson as the project's supervisor, a lasting testament to his original vision, one that infused scientific results with the artistry of illustration.

Shortly before Thomson's death, Murray had taken on the position of director in January 1882 and resolved to reorganise the *Challenger* Office. In 1883 he hired additional office staff, including editorial assistants, and new systems were put in place.[49] For instance, printed illustrations and plates were circulated between artists, printers, authors and editors to ensure that specimens were correctly drawn, printed, labelled and described.[50] Illustrations, like biological specimens, were valuable but could easily go missing in transit. Assistants recorded details of to whom and where plates were sent and, crucially, if the plates had been returned to the Office.[51]

Under Murray's direction and with the aid of this small administrative team, the *Challenger* Office published most of the series before the end of 1889. A few reports remained incomplete, including a summary of the scientific results, authored by Murray and designed as the final volume, which gave detailed lists of the animals obtained by the expedition at different depths, as well as a plethora of charts, maps and a historical introduction to the science of oceanography. The other major work left unfinished by the end of the decade was the *Report on Deep-Sea Deposits*.

One of the benefits of serialisation was that it was possible for scientific research to continue while the first reports were being sent to press. As editor, Murray took advantage of the additional time to expand the project to a scale that dwarfed his preliminary paper of 1876. Continuing their work until 1891, he and Renard analysed not only the *Challenger* materials, but also hundreds of soundings from other expeditions, naval survey vessels and privately owned vessels. Murray explained:

> Since the return of the *Challenger* Expedition very many samples of marine deposits have been collected from nearly all regions of the ocean basins by the surveying vessels of the British Navy, by the telegraph ships belonging to the India Rubber, Gutta-percha and Telegraphs Works Company, and the Telegraph Construction and Maintenance Company, and by Norwegian, Italian, French, German, and American Expeditions.[52]

The additional data rapidly expanded the study's geographical scope — a leap that made understanding the general global distribution of deep-sea sediments possible. Telegraph companies, navies and ocean scientists all shared an interest in mapping the ocean floor.[53] As part of this effort, the Hydrographic Office directed deep-sea sediments collected by Royal Navy survey ships — including the *Bulldog*, *Valorous*, *Nassau*, *Swallow* and *Dove* — to Edinburgh for inclusion in Murray's research. From the American expeditions, Agassiz contributed samples from US Coast Survey ships including

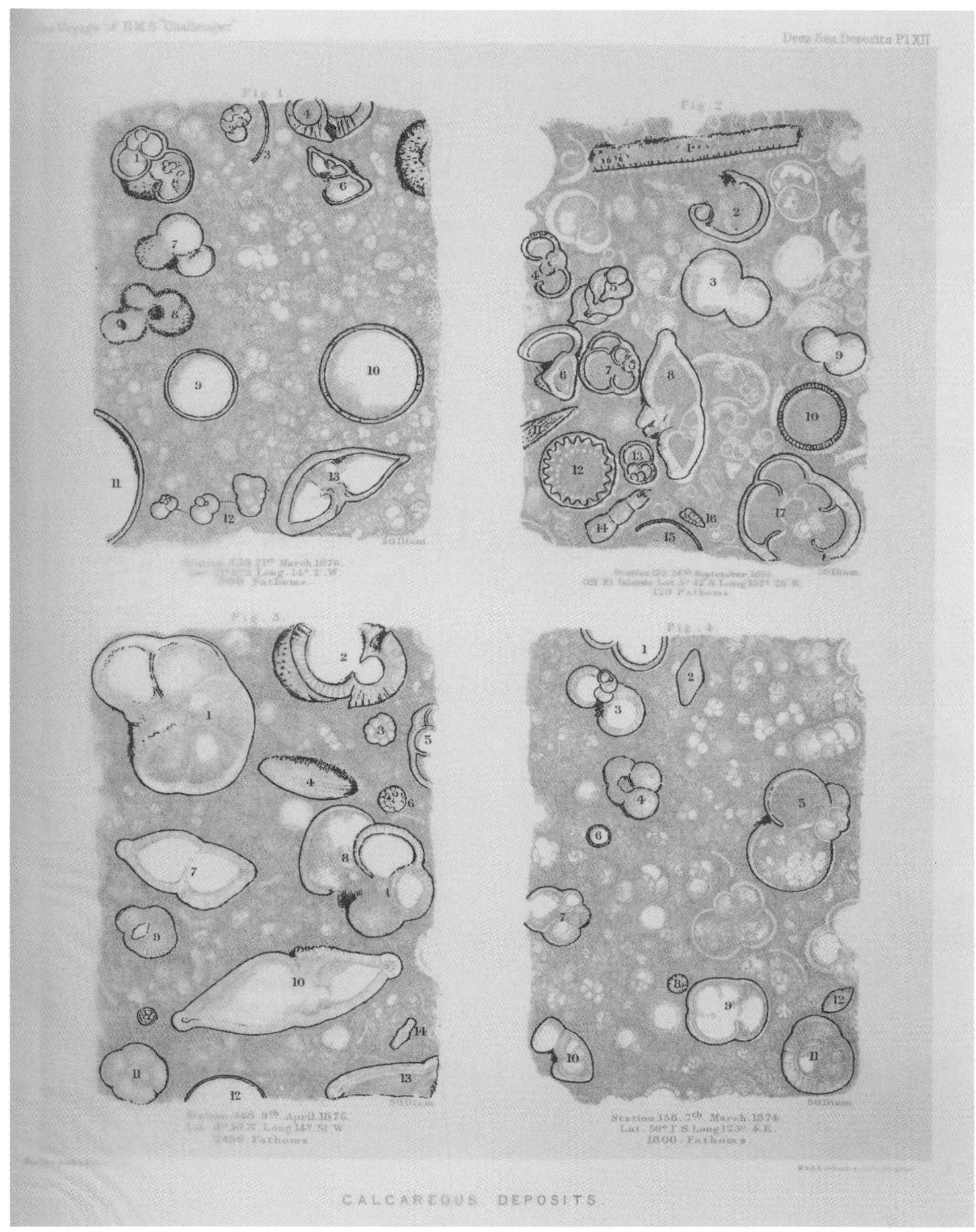

129 A printed tissue overlay outlines species of Foraminifera in calcareous deposits. The same plate without the overlay is shown opposite. The top left figure (fig. 1) is a section of Globigerina ooze from Station 338, from a depth of 1,990 fathoms (3,639 m) in the South Atlantic, magnified 50 diameters. This plate was lithographed by Alphonse-François Renard and George West and probably printed in Brussels.

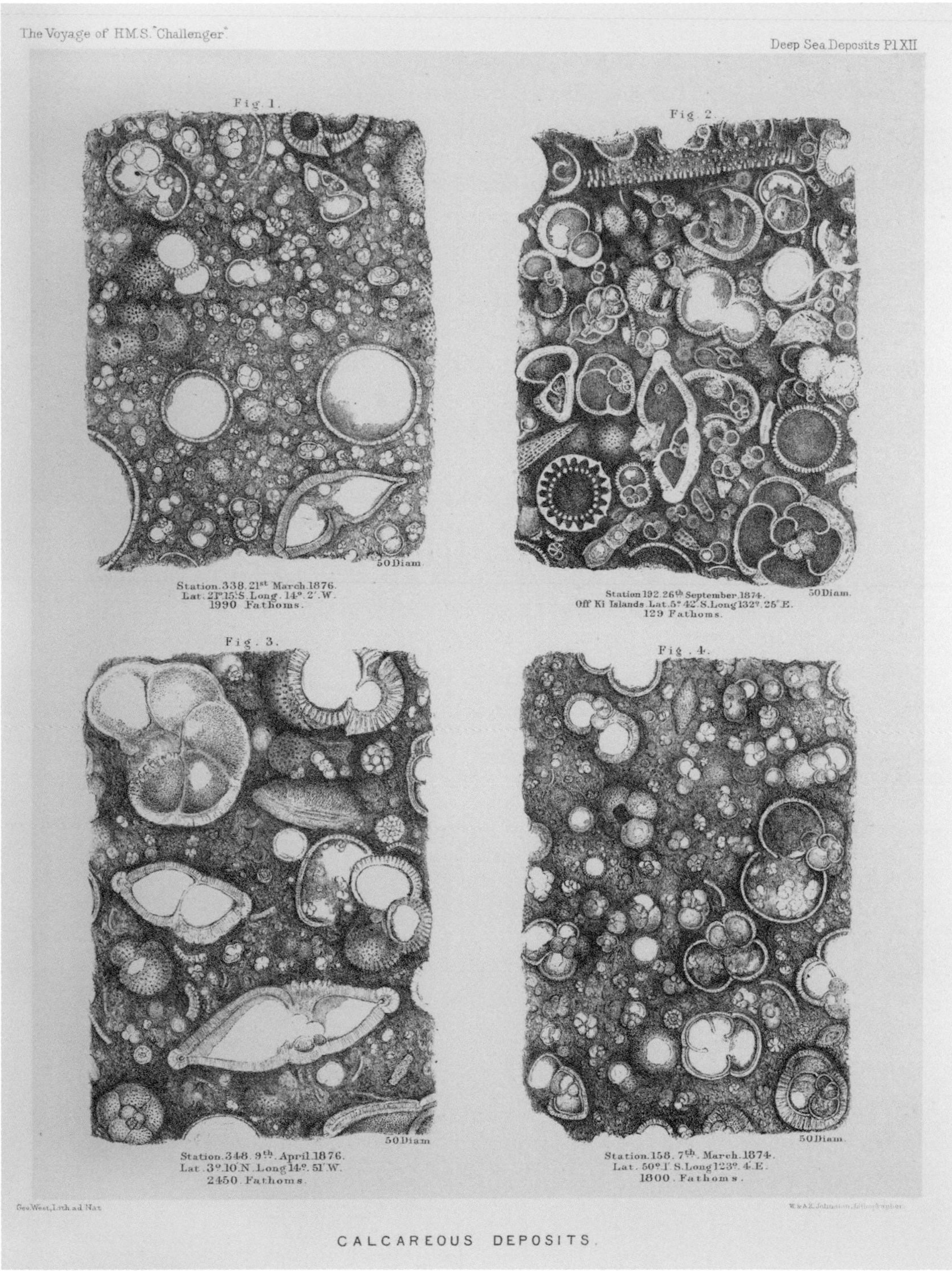

CALCAREOUS DEPOSITS.

the *Blake*, *Tuscarora* and *Gettysburg*.[54] While Murray and Renard's study of marine geology began with samples dredged and sounded by *Challenger*, it morphed into an international collaborative effort centred in Edinburgh.

Typesetters and Printers

Industrial sites on land are not often associated with *Challenger* and the work of oceanographers, but typesetters and printers played an essential role in the making of the expedition's reports. Located in Edinburgh, Neill & Company printed the text and bound most of the *Challenger* series volumes.[55] Since 1783, the company had produced the prestigious *Transactions of the Royal Society of Edinburgh* and had built a reputation serving the intellectuals of the city.[56] It operated printing works at Old Fishmarket, a large industrial building located near the city's harbour.[57] A photograph of the site from the 1880s shows women and men working with steam-powered printing presses (image 130). At the time the *Report on Deep-Sea Deposits* was put together, the company employed some 326 workers, including over 100 female compositors and 19 machine boys and girls.[58]

Close proximity between printer and client was an enormous advantage and Neill & Company operated several offices throughout the city.[59] As an indication of the substantial work of publishing the *Challenger Report*, the firm opened a new office at 32a George Street in 1887, across the street from the Royal Society of Edinburgh and a ten-minute walk from the *Challenger* Office.[60] The reports were not printed as books are today, as a whole, but rather sections were sent off one at a time. The written manuscripts were put into type, and a 'proof' was then printed and returned to the *Challenger* Office for corrections.[61] This led to a constant flow of paper between Queen Street, the printer's offices and industrial works. In some instances, authors (who were located in several countries) had to be consulted. In the final process of manufacture at Old Fishmarket, workers cut and moved paper, set type and operated steam presses.

The Finished Report

While co-authorship had many advantages, it also had its drawbacks. Murray and Renard did not agree on some of the report's conclusions,

130 The industrial works of Neill & Company at Old Fishmarket, Edinburgh, *c.*1880, where thousands of pages of text were printed for the *Challenger* scientific reports.

and they had to find ways to compromise. During the final stages of writing up, Murray wrote to Agassiz in February 1891: 'I am busy with the Deep Sea Deposits: it is a joint book with Renard and I don't find that joint work of that kind progresses rapidly or that it is every way satisfactory in the composition. However I hope it may be out in June at latest.'[62]

After the colour plates were finalised, they were gathered and sent to Neill & Company to be bound with the report's 525 pages of text, 43 charts of *Challenger*'s deep-sea sounding stations and 22 diagrams showing the distribution of temperature and topography across the sections of ocean that the ship traversed. Murray had begun his investigation of marine deposits during the voyage and, after years of study and collaboration, the volume *Report on Deep-Sea Deposits* was finally completed in winter 1891.

One of the book's most consequential achievements was a chart that visualised the global distribution of deep-sea sediments. There is a striking difference between Murray's first map, made from his limited observations during *Challenger*'s 1872–76 voyage (see image 122, pp. 202–03), and the more comprehensive chart published in 1891, which combines the information collected by many other voyages and expeditions (image 131).

The 1891 chart is colour-coded to show the distribution of particular marine deposits, a technique long used by geologists to map classifications of rock on land. The yellow portions designate areas of soft sediment, 'Coral Muds and Sands', found near coral reefs. The pink regions represent 'Globigerina Ooze', sediment of which at least 30 per cent is planktonic Foraminifera (predominantly of the genus *Globigerina*, or marine plankton) and the most

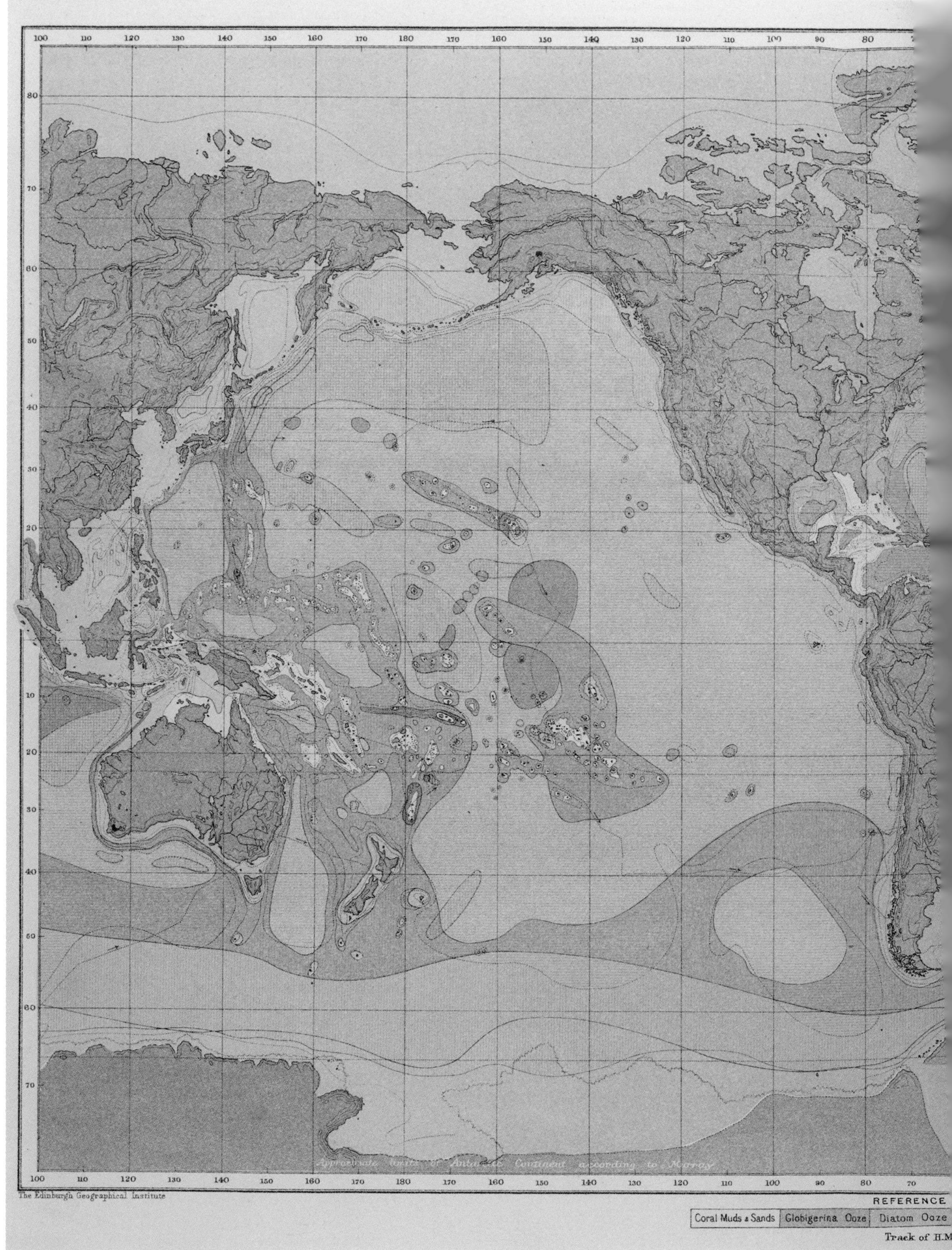

131 Chart showing the global distribution of deep-sea sediments. Published by John Murray and Alphonse-François Renard in 1891.

Deep-Sea Deposits, – CHART 1.
URS OF DEPTH
Approximate limits of Antarctic Continent according to Murray
John Bartholomew & Co.
Red Clay
Terrigenous Deposits (Blue Muds, etc.)

widespread deposit found on the deep-sea floor. Murray and Renard's chart roughly corresponds with current scientific thought that this type of sediment covers most of the western Indian and mid-Atlantic Oceans, and the equatorial and South Pacific.[63] The second most abundant type of deposit, 'Red Clay' (shown in orange) is made of fine clay material derived from the land, transported by winds and ocean currents, and accumulated in the deepest and most remote areas of the ocean. The substance, mostly iron (giving the soil its red colour), aluminium and silica, covers almost half of the Pacific Ocean floor.[64] 'Diatom Ooze', another type of deposit made from the skeletal remains of microscopic floating organisms, is identified in green. Occurring in the Southern Ocean, it contains the frustules (tiny siliceous shells) of diatoms, photosynthesising algae that are a primary food source for higher organisms in the marine food chain.[65]

Combined with the other illustrations in the report, the chart represented a new way of seeing and understanding the ocean floor. For researchers, the visual and descriptive style established in the *Report on Deep-Sea Deposits* offered what was then the most thorough classification scheme for marine deposits, heralding a new era of geological oceanography.

With the publication of the final volume of the *Challenger* series in 1895, the popular and scientific press celebrated and critically reviewed the expedition's published work, now 50 volumes long and with reports on subjects that form pillars of modern oceanography today: marine biology, chemistry and physics, meteorology and marine geology. Thomson's strong vision, founded on the belief that scientific ideas should be accompanied by high-quality illustrations, was vindicated, as the beautiful plates along with numerous woodcuts, charts and diagrams, received wide acclaim and bolstered the report's reputation.

In deciding to include delicate chromolithography, however, the print run was limited to 750 sets, a fact that was criticised in the popular press.[66] The people and institutions most actively engaged in ocean science at the time — members of the expedition, report authors and the country's most prestigious libraries — received the relatively small number of copies. Organisations such as the Hydrographic Office, the University of Edinburgh, the Natural History Museum in London and the Royal Society were given full sets. The remainder were purchased by other individuals, universities, libraries and marine research centres, which furthered the study of oceanography and helped to secure *Challenger*'s international legacy.

The depth and breadth of oceanographic research within the *Challenger Report*, together with the quality printing and production, created enduring reference works that informed future research in Britain and elsewhere. Copies of *Report on Deep-Sea Deposits*, for example, along with the other volumes of the *Challenger Report*, were acquired (either as gifts or purchased) for the libraries of the Zoological Station at Naples, a historic marine research centre founded by Anton Dohrn in 1872, and the Museum of Comparative Zoology at Harvard University, led by Alexander Agassiz. Both institutions maintained leading oceanographic studies departments into the early twentieth century and their libraries made the major published literature available to researchers there. Thanks to the combined efforts of scientists, office assistants, laboratory analysts, artists, lithographers and those who worked at the printing presses, the *Challenger Report* contributed significantly to the development of ocean science and our understanding of the ocean floor today.[68]

The Legacy of the *Challenger* Expedition

In 1895 John Murray published the final two volumes of the *Report on the Scientific Results of the Voyage of H.M.S. Challenger During the Years 1873–76*, marking the culmination of a project that had spanned 23 years. It was a transformative period in the study of the sea. Since the deep-water dredging survey of the *Lightning* in 1868, the field of ocean science had developed from a loose association of naturalists, naval officers and hydrographers to a recognised scientific discipline. In *A Summary of the Scientific Results*, Murray wrote: 'The *Challenger* Expedition has played a very large part in all the recent advances in oceanographical knowledge. The Official Reports on the Scientific Results of the Expedition deal more or less directly with all those branches of knowledge which, we have seen, constitute the science of oceanography.'[1] This book has touched on some of these strands: the physical character of the ocean, including its topography, temperature, chemistry and currents; the study of marine life and deep-sea biology; and the identification of deep-sea deposits and the field of submarine geology.

One hundred and fifty years since the launch of the *Challenger* Expedition, the voyage's bearing on and relevance to modern oceanography is again being celebrated and scrutinised. Although the expedition's scientific work rightly deserves acclaim, *Challenger* was one part in a rich history of efforts to explore the deep sea. To name but a few, at around the same time the United States Coast Survey steamers *Corwin* and *Bibb* (1867–68) accomplished deep-sea dredging off the coast of Florida and the German *Gazelle* Expedition (1874–76) circumnavigated the globe and acquired oceanographic measurements and materials. Leading innovations in deep-sea sounding, the three cruises of the US Coast Survey steamer *Blake* investigated the North Atlantic from 1877 to 1880 with the Sigsbee Sounding Machine, a great improvement on the more cumbersome and heavy hemp rope deployed by *Challenger*'s crew. What, then, distinguished the *Challenger* Expedition from its contemporaries?

At the end of the nineteenth century, oceanography, like the study of Earth's magnetism and astronomy, had become a global enterprise. Thanks to the patronage of the Royal Navy, *Challenger* operated with a geographic reach that had not been possible on previous voyages. The resulting mass of data and specimens collected, dispersed, analysed and circulated in published form after the voyage contributed significantly to our understanding of the ocean not as discrete seas, as had been the case prior to this, but as a connected environment spanning the globe.

The ability to carry out an exploration of the deep sea on such a monumental scale is thus at the heart of the *Challenger* narrative and the development of oceanography. In part, the scientific programme was made viable by nineteenth-century revolutions in how people and things moved. Transport and communication systems dating from the 1840s and 1850s came to define modern society; these same forces were essential to early investigations of the ocean's depths.[2] The national post, telegraphic cables, railways, photography, steam-powered publishing and scheduled steamship services all facilitated *Challenger*'s mission at sea and the transmission of information back to Britain.

During this new age, goods were being shipped across oceans and continents at faster speeds and with increased reliability. Hundreds of boxes and barrels of marine specimens were conveyed to London from major ports using the Royal Mail, a recently established service heavily subsidised by the British government. By the end of the voyage, the *Challenger* Expedition had assembled the largest assortment of deep-sea animals to date. The collection proved the abundance and variety of marine life throughout the oceans. The expedition's readings, measurements and records also created a valuable historical benchmark that climate change scientists still refer to today.

While the personalities of Thomson, Murray and Nares have been credited historically with the success of the voyage, the *Challenger* Expedition needs to be considered within its wider context, acknowledging the resources and influence of the British Empire. With the backing of the government, the project also benefited from the speed, manpower and reserves of the Royal Navy, the nineteenth century's dominant naval force. To respond to events and changing priorities across the Americas, Asia and Africa, the fleet was supported by a chain of dry docks, naval bases and coaling stations. This same infrastructure enabled *Challenger* to conduct steam-powered oceanographic experiments and sustain a long voyage throughout the Atlantic, Pacific and Southern Oceans, a feat unmatched by other nations (image 132).

British politics, commerce and science supported the project, but these factors also shaped its scientific results. *Challenger*'s sounding stations were located not at evenly spaced points throughout the voyage, as has often been assumed, but in areas of interest to the Hydrographic Office and telegraph companies. Distributed over a wide geographical area, they began to reveal the hidden topography of Earth's ocean basins for the first time. Exploring the deep ocean remains a costly endeavour, funded and directed more and more by national and commercial interests, such as military hydrography, offshore oil drilling and the growing field of deep-sea mining.

The examination of the extensive *Challenger* data, deep-sea deposits and natural history collections fostered multinational collaboration in the study of the sea, a defining quality of present-day oceanography. After the voyage concluded, the work of analysis involved dozens of experts located in several countries. The trials and tribulations of Alexander Agassiz demonstrated how marine biology in the late nineteenth century, while becoming an increasingly professional field, often necessitated personal wealth and social connections. Agassiz helped assign *Challenger* specimens to naturalists located in Britain, Europe and the United States, several of whom he and Thomson had met and corresponded with as part of their previous studies and travels.

Both the scientific content and design of the serialised *Report on the Scientific Results of the Voyage of H.M.S. Challenger* assisted in establishing *Challenger*'s enduring legacy. Artists, illustrators, analysts, naturalists and printers worked together to create the durable and beautiful volumes. Given as gifts or purchased, the 750 sets found their way to marine research centres, libraries, universities and natural history museums. Providing a framework for subsequent studies, the reports combined findings gathered from many voyages and encompassed subjects that came to be known collectively as 'oceanography' by the end of the century.

The use of photography as part of the expedition made the ocean visible to scientists and the public in new ways; the first photos of icebergs in the Antarctic are a notable example. With the benefit of specialised rooms and equipment on board, images were copied and disseminated for use in publications and purchased by the crew. Organised by Captain William de Wiveleslie Abney, photography was largely conducted in the manner of the Royal Engineers, but the three official photographers took varying approaches to recording the landscape and documenting members of the expedition and people they encountered.

Challenger was a trailblazer and our capacity for understanding the ocean has since been extended by technological advances. Researchers now deploy submersibles, sonar, high-tech buoys and remotely operated vehicles to probe deep-sea environments. Over the past decade, high-resolution underwater video cameras have enabled scientists and the public to view marine life as never before. Oceanographic research continues as an international project that calls for ships at sea and laboratories on land, but also relies on less conspicuous labours and resources, including harbour facilities, electronics factories and satellite navigation systems maintained by the United States, Russia, the European Union and China.

As we strive to understand our blue planet, a more inclusive history of ocean science is relevant to how the ocean is studied today. Everywhere *Challenger* travelled, the expedition acquired knowledge from those who lived in tandem with the sea or along the coast. The innovations and practices of Indigenous Peoples and local communities are of fundamental importance in informing strategies for conservation and sustainable, equitable use of marine resources and are increasingly recognised as such.[3]

We have reached a crucial moment in our history. The unprecedented decline of biodiversity and marine habitats, caused by growing pressures from human activity, pollution and climate change, has become the most pressing environmental concern of our time. Whole ecosystems are on the verge of destruction.[4] Never before have we had such an appreciation for and insight into the bounty of life and mosaic of forces that constitute the ocean, a realm once beyond human observation. Today, our greatest challenge is to act upon this knowledge — to support the critical efforts being made to protect the vast ocean ecosystem and ensure the future well-being of humanity and all life on Earth.

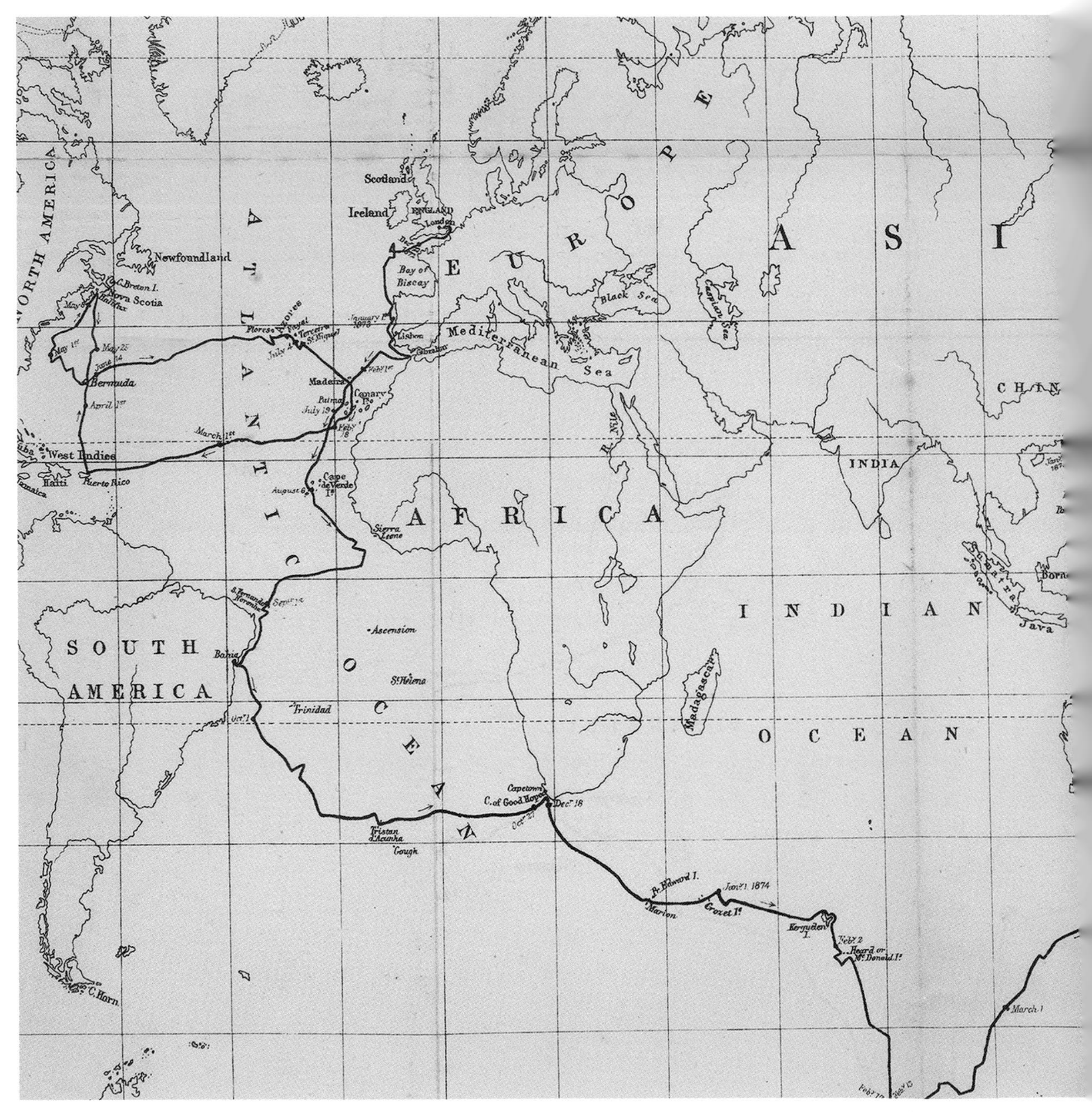

132 Map of the route of HMS *Challenger* between December 1872 and June 1876.

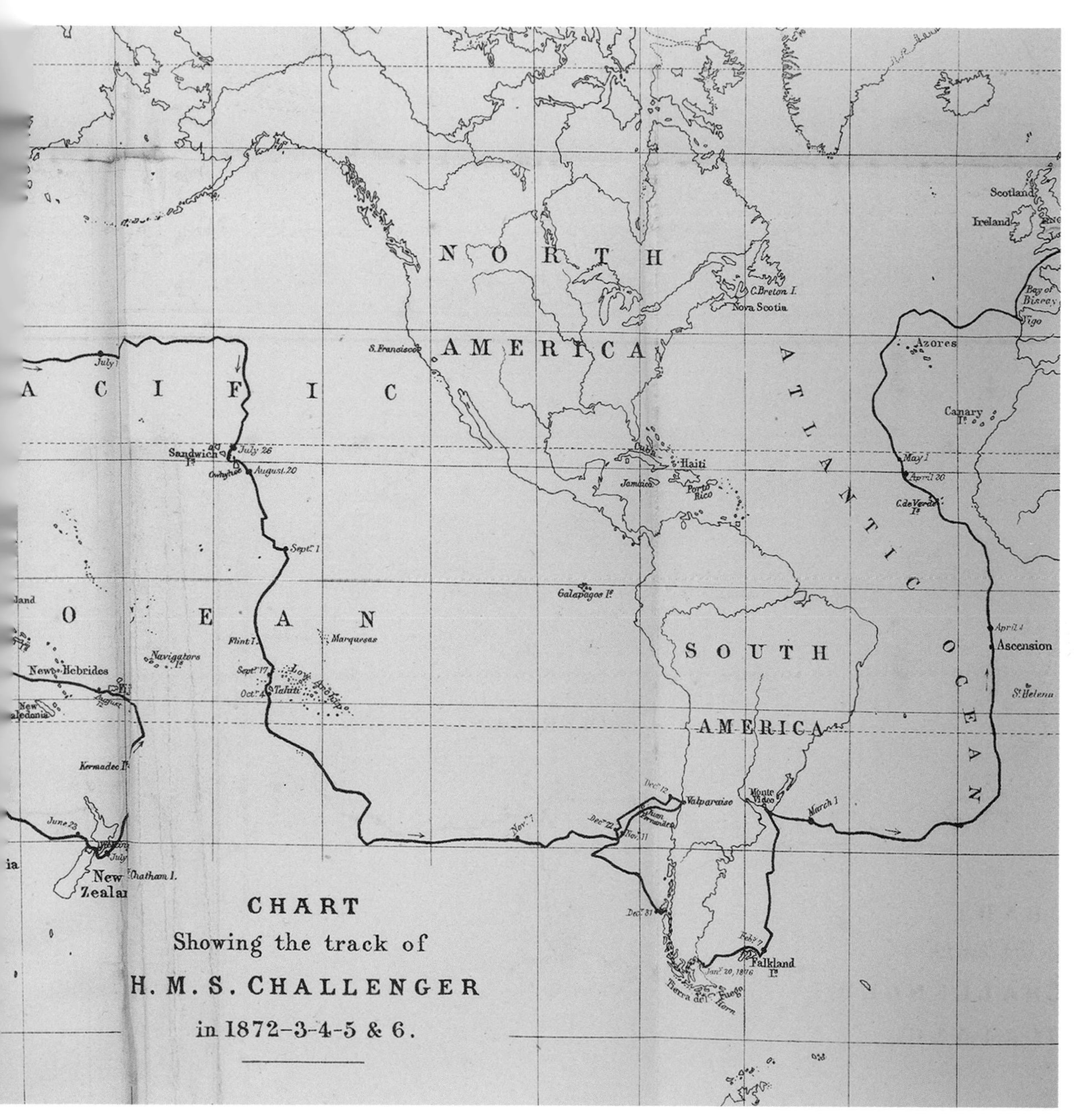
CHART
Showing the track of
H. M. S. CHALLENGER
in 1872-3-4-5 & 6.
NORTH
AMERICA
SOUTH
AMERICA
ATLANTIC
OCEAN
Scotland
Ireland
Bay of Biscay
Vigo
Azores
Canary Is.
C. de Verde Is.
Ascension
St. Helena
C. Breton I.
Nova Scotia
S. Francisco
Cuba
Haiti
Jamaica
Porto Rico
Galapagos Is.
Valparaiso
Monte Video
Falkland Is.
Tierra del Fuego
Juan Fernandez
Sandwich Is.
Owyhee
July 26
August 20
Sept. 1
Flint I.
Marquesas
Navigators Is.
Tahiti
New Hebrides
Kermadec Is.
Chatham I.
New Zealand
March 1
April 30
May 1
Nov. 7
Nov. 11
Dec. 12
Dec. 31
Jan. 20, 1876

The Route of HMS *Challenger*, December 1872–June 1876

From	To	Departure	Arrival	Distance travelled
Sheerness	Portsmouth	7 December 1872	11 December 1872	200
Portsmouth	Lisbon	21 December 1872	3 January 1873	1,091
Lisbon	Gibraltar	12 January 1873	18 January 1873	340
Gibraltar	Madeira	26 January 1873	3 February 1873	655
Madeira	Tenerife	5 February 1873	7 February 1873	255
		(Off Tenerife)		230
Tenerife	St Thomas	14 February 1873	16 March 1873	2,879
St Thomas	Bermuda	24 March 1873	4 April 1873	870
Bermuda	Halifax, via New York	21 April 1873	9 May 1873	1,261
Halifax	Bermuda	19 May 1873	31 May 1873	796
Bermuda	São Miguel, Azores	13 June 1873	4 July 1873	2,031
São Miguell	Madeira	9 July 1873	16 July 1873	528
Madeira	São Vicente	17 July 1873	27 July 1873	1,066
São Vicente	Praia	5 August 1873	7 August 1873	170
Praia	St Paul Rocks	9 August 1873	27 August 1873	1,955
St Paul Rocks	Fernando de Noronha	29 August 1873	1 September 1873	342
Fernando de Noronha	Bahia	3 September 1873	14 September 1873	815
Bahia	Cape of Good Hope	25 September 1873	28 October 1873	3,883
Cape of Good Hope	Melbourne	17 December 1873	17 March 1874	7,637
Melbourne	Sydney	1 April 1874	6 April 1874	550
Sydney	Wellington	8 June 1874	28 June 1874	1,432
Wellington	Tongatapu	7 July 1874	19 July 1874	1,547
Tongatapu	Galoa Harbour, Kadavu Island	22 July 1874	25 July 1874	400
Galoa Harbour	Levuka	27 July 1874	28 July 1874	120
Levuka	Galoa Harbour	1 August 1874	3 August 1874	120

Coal burnt	No. of days at sea		No. of deep-sea soundings obtained	No. of temperature readings taken	No. of successful dredgings	No. of successful trawlings
Tons	Days	Hours				
87	5					
207	13		4		1	
68	7		10		3	1
100	9		12			3
15	2		1			
45	4		11	2	2	
122	30		24	13	11	2
79	11		10	5	6	
127	18		14	7	9	
158	12		11	8	7	2
109	21		18	12	4	6
34	7		6	4	3	
46	10		11	7	3	1
12	2		1			
101	18		12	12	1	4
18	3		5	2		
87	11		17	2	2	7
173	33		13	10	9	2
247	91		15	13	11	6
58	5		1	1	4	3
177	20		11	6	1	5
73	13		6	4		5
13	3				2	1
7	1					
33	2		1	1	3	

From	To	Departure	Arrival	Distance travelled
Galoa Harbour	Pabaju (Albany Island)	10 August 1874	1 September 1874	2,250
Pabaju (Albany Island)	Dobo, Aru Islands	8 September 1874	16 September 1874	656
Dobo	Kai Islands	23 September 1874	24 September 1874	100
Kai Islands	Banda Islands	26 September 1874	29 September 1874	200
Banda Islands	Ambon Island	2 October 1874	4 October 1874	115
Ambon Island	Ternate	10 October 1874	14 October 1874	300
Ternate	Zamboanga	17 October 1874	23 October 1874	511
Zamboanga	Iloilo	26 October 1874	28 October 18741	220
Iloilo	Manila	31 October 1874	4 November 1874	350
Manila	Hong Kong	11 November 1874	16 November 1874	650
Hong Kong	Manila	6 January 1875	11 January 1875	650
Manila	Cebu	14 January 1875	18 January 1875	380
Cebu	Camiguin Island	24 January 1875	26 January 1875	110
Camiguin Island	Zamboanga	26 January 1875	29 January 1875	250
Zamboanga	Humboldt Bay	5 February 1875	23 February 1875	1,333
Humboldt Bay	Admiralty Islands	24 February 1875	3 March 1875	403
Admiralty Islands	Yokohama	10 March 1875	11 April 1875	2,533
Yokohama	Kobe	11 May 1875	15 May 1875	350
Kobe	Mihara	25 May 1875	26 May 1875	120
Mihara	Kobe	28 May 1875	29 May 1875	120
Kobe	Yokohama	2 June 1875	5 June 1875	400
Yokohama	Honolulu	16 June 1875	27 July 1875	4,302
Honolulu	Hilo	1 August 1875	4 August 1875	200

Coal burnt	No. of days at sea		No. of deep-sea soundings obtained	No. of temperature readings taken	No. of successful dredgings	No. of successful trawlings
	Days	Hours				
71	22		9	8	5	3
24	8				5	3
17	1		2	2		1
38	3		1	1		4
17	2					
30	4		2	2		1
48	6		2	1	2	2
21	2		1	1		1
38	4				4	
24	5		1	1		1
35	5		1	1		
45	4		1	1		4
	2		1	1	1	1
	3		1	1		
108	18		6	5		5
42	7		1	1		1
106	32		13	12	2	3
72	4		1	1	1	1
20	1					2
19	1					2
80	3		4	3		2
279	42		24	24	2	7
60	3		1	1		

From	To	Departure	Arrival	Distance travelled
Hilo	Tahiti	19 August 1875	18 September 1875	2,630
Tahiti	Juan Fernández Islands	3 October 1875	13 November 1875	4,643
Juan Fernández Islands	Valparaíso	15 November 1875	19 November 1875	400
Valparaíso	Gulf of Penas	11 December 1875	31 December 1875	2,033
Gulf of Penas	off Isla Wager	1 January 1876	1 January 1876	
off Isla Wager	Puerto Gray	2 January 1876	2 January 1876	
Puerto Gray	Puerto Grappler	4 January 1876	4 January 1876	
Puerto Grappler	Tom Bay	5 January 1876	5 January 1876	
Tom Bay	Puerto Bueno	8 January 1876	8 January 1876	700 (cumulative, 1—18 Jan.)
Puerto Bueno	off Isla Emiliano Figueroa	10 January 1876	10 January 1876	
Bahía Isthmus	Puerto Churruca	11 January 1876	11 January 1876	
Puerto Churruca	Puerto del Hambre	13 January 1876	13 January 1876	
Puerto del Hambre	Punta Arenas	14 January 1876	14 January 1876	
Punta Arenas	Isabel Island	18 January 1876	18 January 1876	
Isabel Island	Falkland Islands	20 January 1876	22 January 1876	400
Falkland Islands	Montevideo	6 February 1876	15 February 1876	1,173
Montevideo	Ascension Island	25 February 1876	27 March 1876	3,729
Ascension Island	Praia	3 April 1876	17 April 1876	1,620
Praia	Porto Grande	17 April 1876	18 April 1876	180
Porto Grande	Vigo	26 April 1876	20 May 1876	2,926
Vigo	Portsmouth	21 May 1876	24 May 1876	630
Portsmouth	Sheerness	25 May 1876	26 May 1876	150
			Grand total	68,890

Coal burnt	No. of days at sea		No. of deep-sea soundings obtained	No. of temperature readings taken	No. of successful dredgings	No. of successful trawlings
	Days	Hours				
189	30		17	17	1	5
222	41		22	19		11
17	4		1	1		1
76	21		4	5	1	4
9	0	14	1	1		1
7	0	13 ½	1			1
6	0	1				
9	0	12 ½	1			1
10	0	11 ½	1			1
13	0	12 ½	1			1
10	0	10	1			1
14	0	13 ½				
5	0	4				
6	0	3 ½				
52	3	0	3	1		3
80	9	0	4	4		2
177	31	0	21	20	3	4
187	14	0	3	8	2	
23	1	0				
141	24	0	2	2		
122	3	0				
30	1	0				
4,826	719	0 ¼	374	255	111	129

NOTES

GENERAL NOTE ON REFERENCES

Volumes prepared by Charles Wyville Thomson and John Murray that form part of the series of reports published by the *Challenger* Office are given by volume title and indicated as belonging to the official *Challenger Report*. The complete title of the series is *Report on the Scientific Results of the Voyage of H.M.S. Challenger During the Years 1873–76*, 50 vols (Edinburgh: HMSO, 1880–95).

INTRODUCTION

1 For an overview of nineteenth-century ocean science, see Helen Rozwadowski, *Fathoming the Ocean* (Cambridge, MA, and London: Belknap Press, 2005); Margaret Deacon, *Scientists and the Sea, 1650–1900: A Study of Marine Science* (Aldershot, Hampshire: Ashgate, 1997), pp. 276–332; Susan Schlee, *A History of Oceanography* (London: Robert Hale, 1973), pp. 23–189; Ian Jones and Joyce Jones, *Oceanography in the Days of Sail* (Sydney: Hale & Iremonger, 1992), pp. 99–246.

2 *Challenger* conducted oceanographic experiments at 354 stations, yet more are detailed in the scientific reports. For instance, there is a station 166 but the report also lists stations 166A, 166B and 166C. Taking these additional geographical points into account, *Challenger* carried out various experiments at a total of 504 stations. Sounding was carried out at all stations and the latitude and longitude recorded. Thomas Henry Tizard, Henry Nottidge Moseley, John Young Buchanan and John Murray, *Narrative of the Cruise of H.M.S. Challenger, with a General Account of the Scientific Results of the Expedition* [*Challenger Report*], 2 vols (Edinburgh: HMSO, 1885), vol. 1, Appendix II, pp. 1007–15.

3 This is not the total number of experiments and soundings carried out during the voyage but those considered to be 'deep-sea' investigations. William J.J. Spry, *The Cruise of Her Majesty's Ship 'Challenger'* (London: Sampson Low, Marston, Searle and Rivington, 1876), Appendix, pp. 316–19.

CHAPTER 1

1 Charles Wyville Thomson, *The Depths of the Sea: an account of the general results of the dredging cruises of H.M.S.s 'Porcupine' and 'Lightning' during the summers of 1868, 1869, and 1870, under the scientific direction of Dr. Carpenter, F.R.S., J. Gwyn Jeffreys, F.R.S., and Dr. Wyville Thomson, F.R.S.* (London: Macmillan, 1873), pp. 1–2.

2 'John Ross first North-West Passage expedition 1818', Royal Museums Greenwich, www.rmg.co.uk/stories/topics/john-ross-first-north-west-passage-expedition-1818 (accessed 27 July 2022).

3 Cara Giaimo, 'How a Fake Mountain Slowed Down Arctic Exploration', Atlas Obscura, www.atlasobscura.com/articles/john-ross-arctic-mountain-fake (accessed 27 July 2022).

4 John Ross, *A Voyage of Discovery, Made under the Orders of the Admiralty, in His Majesty's Ships Isabella and Alexander for the Purpose of Exploring Baffin's Bay, and Inquiring into the Possibility of a Northwest Passage*, 2 vols (London: John Murray, 1819), vol. 2, appendix no. XIII, p. 247.

5 Ibid., p. 178.

6 Thomson, *Depths of the Sea*, pp. 18–20.

7 J. Ross, *Voyage of Discovery*, vol. 2, p. 153.

8 John Ross, *Narrative of a Second Voyage in Search of a North-West Passage*, 2 vols (London: A.W. Webster, 1835), vol. 1, p. 555.

9 Ibid., Natural History Appendix, Plate B.

10 See James Clark Ross, *A Voyage of Discovery and Research in the Southern and Antarctic Regions, during the Years 1839–43* (London: John Murray, 1847).

11 Deacon, *Scientists and the Sea*, p. 282.

12 J.C. Ross, *Voyage of Discovery and Research*, pp. 201–02.

13 Ibid.

14 E. Janet Browne, *Charles Darwin: Volume 1, Voyaging* (London: Jonathan Cape, 1995), pp. 148–49.

15 'Letter no. 105, J.S. Henslow to Charles Darwin, 24 August 1831, Cambridge', *Darwin Correspondence Project*, www.darwinproject.ac.uk/letter/?docId=letters/DCP-LETT-105.xml (accessed 24 June 2022).

16 'Letter no. 106, George Peacock to Charles Darwin, [*c.* 26 August 1831]', ibid. (accessed 27 April 2022).

17 Charles Darwin, 'Beagle diary', pp. 399–400; 'Darwin & coral reefs', ibid. (accessed 23 November 2021).

18 Edward Forbes, *Literary Gazette*, 7 November 1840, pp. 725–26.

19 Rozwadowski, *Fathoming the Ocean*, pp. 128–31.

20 Ibid., pp. 131–32.

21 Edward Forbes and Robert Godwin-Austen (eds), *The Natural History of the European Seas* (London: John Van Voorst, 1859), p. 26.

22 Ibid., pp. 26–27.

23 See Joseph-Fortuné-Théodose Eydoux and Louis-François-Auguste Souleyet, *Voyage autour du monde exécuté pendant les années 1836 et 1837 sur la corvette la Bonite, commandée par M. Vaillant. Histoire naturelle. Zoologie. Atlas* (Paris: Arthus Bertrand, 1851).

24 Jones and Jones, *Oceanography in the Days of Sail*, p. 125.

25 Ibid., pp. 127, 129; also see Urbain Dortet de Tessan, *Voyage autour du monde sur la frégate la Vénus, pendant les annees 1836–1839. Physique*, 5 vols (Paris: Gide, 1842–44).

26 William Stanton, *The Great United States Exploring Expedition of 1838–1842* (Berkeley: University of California Press, 1975), pp. 205–15.

27 For more on European exploration, Indigenous agency and the production of knowledge in the Pacific, see Shino Konishi, Maria Nugent and Tiffany Shellam (eds), *Indigenous Intermediaries: New Perspectives on Exploration Archives* (Canberra: ANU Press, 2015).

28 Smithsonian Libraries, Nathaniel Philbrick, 'The United States Exploring Expedition, 1838–1842', January 2004, https://www.sil.si.edu/DigitalCollections/usexex/learn/Philbrick.htm (accessed 23 November 2021).

29 Rozwadowski, *Fathoming the Ocean*, p. 50.

30 Smithsonian Institution Archives, 'Record Unit 7253: United States North Pacific Exploring Expedition (1853–1856), Collection Overview', https://siarchives.si.edu/collections/siris_arc_217410 (accessed 16 May 2022).

31 Rozwadowski, *Fathoming the Ocean*, p. 50.

32 Ibid., pp. 52–54.

33 Letter submitted to the government, 23 July 1845, from J.W. and J. Brett to the Right Honourable Sir Robert Peel, printed by Brett's Electric Telegraph, History of the Atlantic Cable & Undersea Communications, 'Jacob Brett, 1807–1897', https://atlantic-cable.com/CablePioneers/JBrett.htm (accessed 23 November 2021).

34 Ibid.
35 National Army Museum, 'Field telegraph wire used during the Crimean War, 1855', NAM.1965-10-202-4, https://collection.nam.ac.uk/detail.php?acc=1965-10-202-4 (accessed 23 November 2021).
36 Matthew Fontaine Maury, *The Physical Geography of the Sea* (London: Sampson Low, Son & Co., 1855).
37 Cyrus West Field, 'The Atlantic cable', *Scientific American*, 21 August 1858, p. 398.
38 Allison Marsh, 'The first transatlantic telegraph cable was a bold, beautiful failure', *IEEE Spectrum*, 31 October 2019, https://spectrum.ieee.org/the-first-transatlantic-telegraph-cable-was-a-bold-beautiful-failure (accessed 23 November 2021).
39 'Report of the Submarine Telegraph Committee', *Parliamentary Papers*, 62, 1860, p. xxxv. Cited in Crosbie Smith and M. Norton Wise, *Energy and Empire: A Biographical Study of Lord Kelvin* (Cambridge: Cambridge University Press, 1989), p. 740. Also see Ben Marsden and Crosbie Smith, *Engineering Empires: A Cultural History of Technology in Nineteenth-Century Britain* (Basingstoke: Palgrave Macmillan, 2005), pp. 178–225.
40 Penelope Hardy and Helen Rozwadowski, 'Maury for modern times: navigating a racist legacy in ocean science', *Oceanography* 33:3 (2020), pp. 10–15, p. 10.
41 Maury, *The Physical Geography of the Sea.*
42 Hardy and Rozwadowski, 'Maury for modern times', pp. 10, 13–14.
43 Bill Glover, 'Great Eastern', History of the Atlantic Cable & Submarine Telegraphy, https://atlantic-cable.com/Cableships/GreatEastern/index.htm (accessed 27 July 2022).
44 John Munro, *Heroes of the Telegraph* (London: Religious Tract Society, 1891), p. 46.
45 Thomson, *Depths of the Sea*, p. 28.
46 Erik Ducker, *News from an Inaccessible World: The History and Present Challenges of Deep-Sea Biology* (Enschede: Gildeprint, 2014), pp. 40–41.
47 Rozwadowski, *Fathoming the Ocean*, p. 148.
48 Ducker, *News from an Inaccessible World*, p. 41.
49 Thomson, *Depths of the Sea*, p. 49.
50 Ibid., p. 50.
51 Deacon, *Scientists and the Sea*, p. 297.
52 Thomson, *Depths of the Sea*, pp. 66–68.
53 Ibid., pp. 60, 69.
54 Ibid., p. 79.
55 Ibid., p. 84.
56 Charles Wyville Thomson (*Nature*, 20 March 1873), cited in Thomas H. Huxley, *Collected Essays*, 9 vols, 'The problems of the deep sea', *Discourses: Biological & Geological*, vol. 8 (London: Macmillan, 1894), p. 44.
57 Huxley, *Collected Essays*, vol. 8, p. 43.
58 Thomson (*Nature*, 20 March 1873), cited in Huxley, *Collected Essays*, vol. 8, p. 44.

CHAPTER 2

1 John Murray and Johan Hjort, *The Depths of the Ocean: A General Account of the Modern Science of Oceanography Based Largely on the Scientific Researches of the Norwegian Steamer Michael Sars in the North Atlantic* (London: Macmillan, 1912), p. 22.
2 Tizard et al., *Narrative of the Cruise*, vol. 1, p. L.
3 Charles Wyville Thomson, *The Voyage of the 'Challenger': The Atlantic; a Preliminary Account of the General Results of the Exploring Voyage of H.M.S. 'Challenger' During the Year 1873 and the Early Part of the Year 1876*, 2 vols (London: Macmillan, 1877), vol. 1, p. 73.
4 Ibid.
5 Ibid., p. 74.
6 Ibid., p. 23.
7 'Report of the Committee appointed at the Meeting of the Council, 26 October, to consider the Scheme of a Scientific Circumnavigation Expedition', Minutes of Council of the Royal Society, 30 November 1871, in Thomson, *Voyage of the 'Challenger': The Atlantic*, vol. 1, pp. 75–77.
8 Letter from the Royal Society to the Secretary of the Admiralty, 8 December 1871, in ibid., pp. 77–78.
9 Deacon, *Scientists and the Sea*, pp. 331–35.
10 Rif Winfield, *British Warships in the Age of Sail 1817–1863: Design, Construction, Careers and Fates* (Barnsley: Seaforth, 2014), p. 192; Tony Rice, *British Oceanographic Vessels 1800–1950* (London: Ray Society, 1986). pp. 30–39.
11 John F. Beeler, *British Naval Policy in the Gladstone-Disraeli Era, 1866–1880* (Stanford, CA: Stanford University Press, 1997), p. 37.
12 'Fiji', *Sydney Mail*, 19 September 1868, p. 11 (trove.nla.gov.au, accessed 26 July 2022).
13 'British paddle steamer Lightning (1829)', https://threedecks.org/index.php?display_type=show_ship&id=21621 (accessed 6 April 2019).
14 Letter from the Admiralty to the Secretary of the Royal Society, 22 August 1872, in Thomson, *The Voyage of the 'Challenger': The Atlantic*, vol. 1, p. 80.
15 See 'Vice-Admiral Sir George Nares, K.C.B., F.R.S.', *Nature* 94 (1915), pp. 565–67. Also, Dartmouth College Library, 'Sir George Strong Nares (1831–1915)', *Encyclopedia Arctica 15: Biographies*, https://collections.dartmouth.edu/arctica-beta/html/EA15-52.html (accessed 23 May 2022).
16 Penelope K. Hardy, 'Science from the quarterdeck: naval-scientific networks on the 1870s *Challenger* Expedition', paper presented at the *26th International Congress of History of Science and Technology* (virtual), Prague, Czech Republic, 27 July 2021.
17 Nares to Richards, 20 October 1871; Box 9: Surveyors Letters 1871; S Papers; United Kingdom Hydrographic Office. Quoted by Hardy, 'Science from the quarterdeck'.
18 Thomson, *Depths of the Sea.*
19 Rozwadowski, *Fathoming the Ocean*, pp. 183–85, 187.
20 Thomson, *Depths of the Sea*, p. 2.
21 Ibid., p. 60.
22 Ibid., p. 5.
23 Ibid., p. 30.
24 Ibid., p. 28.
25 Doug Macdougall, *Endless Novelties of Extraordinary Interest: The Voyage of H.M.S. Challenger and the Birth of Modern Oceanography* (New Haven, CT, and London: Yale University Press, 2019), pp. 33–34.
26 Thomson, *Depths of the Sea*, p. 11.
27 Ibid., p. 18.
28 *Memoirs of Abraham Smith, RN, 1859–1880*, National Maritime Museum, Greenwich, London, BGR/41, p. 101.
29 Anthony Adler, 'The ship as laboratory: making space for field science at sea', *Journal of the History of Biology*, 47, 2014, pp. 333–62, p. 351.
30 Thomas Henry Tizard, 'Hydrography and navigation', *Natural Science: A Monthly Review of Scientific Progress* 41:7 (1895), pp. 10–14, p. 10.
31 Murray and Hjort, *The Depths of the Ocean*, p. 22.
32 Philip Pearson, *A Challenger's Song* (London: Austin Macauley Publishers, 2021), pp. 11, 120.

33 John Colgate Hoyt (ed.), 'The Stoker', in *Old Ocean's Ferry: The Log of the Modern Mariner, the Trans-Atlantic Traveler, and Quaint Facts of Neptune's Realm* (New York: Bonnell, Silver and Co., 1900), p. 121.
34 Letter from Joseph Matkin to his cousin 'Tom' John Thomas Swann, 17 December 1872, in Philip F. Rehbock (ed.), *At Sea with the Scientifics: The Challenger Letters of Joseph Matkin* (Honolulu: University of Hawaii Press, 1992), p. 23.
35 Ibid.
36 'Abstract of the Voyage of HMS *Challenger*', in Spry, *The Cruise of Her Majesty's Ship 'Challenger'*, pp. 387–88.
37 Letter from Matkin to his mother, 30 December 1872, off Vigo, in Rehbock (ed.), *At Sea with the Scientifics*, p. 30.
38 Tizard et al., *Narrative of the Cruise*, vol. 1, p. 185.
39 Letter from Matkin to his mother, 24 February 1875, Humboldt Bay, New Guinea, in Rehbock (ed.), *At Sea with the Scientifics*, p. 228.
40 Letter from Matkin to Swann, 21 September 1873, Bay of Bahia, in ibid., p. 102.
41 Spry, *The Cruise of Her Majesty's Ship 'Challenger'*, p. 21.
42 Letter from Matkin to Swann, 18 January 1873, Gibraltar, in Rehbock (ed.), *At Sea with the Scientifics*, p. 42.
43 Letter from Matkin to his mother, 8 January 1873, Lisbon, in ibid., p. 37.
44 Ibid., p. 37.
45 Letter from Matkin to his mother, Sydney, n.d., in ibid., p. 162.
46 *Memoirs of Abraham Smith*, National Maritime Museum (NMM), Greenwich, London, BGR/41, pp. 104–05.
47 'Abstract of the Voyage of HMS *Challenger*', in Spry, *The Cruise of Her Majesty's Ship 'Challenger'*, pp. 387–88.
48 Letter from Matkin to his mother, Vigo, 30 December 1872, in Rehbock (ed.), *At Sea with the Scientifics*, pp. 30–31.
49 Letter from Matkin to his mother, March 1873, in ibid., p. 56.
50 Ibid., p. 372.
51 Tizard et al., *Narrative of the Cruise*, vol. 1, Appendix IV, pp. 1027–31.
52 Rehbock (ed.), *At Sea with the Scientifics*, p. 371.
53 Letter from Matkin to his mother, 19 December 1874, in ibid., p. 212.
54 Letter from Matkin to his mother, New Guinea, 24 February 1875, in ibid., p. 224.
55 Ibid., pp. 371–72.
56 Letter from Matkin to his mother, Sydney, March 1873, in ibid., pp. 162–63.
57 *Memoirs of Abraham Smith*, NMM, BGR/41, pp. 114–15.

CHAPTER 3

1 Murray and Hjort, *Depths of the Ocean*, p. 22.
2 Tizard et al., *Narrative of the Cruise*, vol. 1, Appendix B, 'Hydrographic instructions to Captain G.S. Nares, HMS *Challenger*', p. 34.
3 Ibid.
4 Ibid., vol. 1, Appendix A, 'Official Correspondence with reference to the *Challenger* Expedition. Extracted from the Minutes of Council of the Royal Society, 20 June 1872', p. 23.
5 Thomson, *Depths of the Sea*, p. 205.
6 For sounding with a lead, see R.W. Cooper, 'Heaving the lead', *Journal of Navigation* 63 (2010), 183–86.
7 Tony Rice, 'The *Challenger* Expedition: the end of an era or a new beginning?', in Margaret Deacon, Tony Rice and Colin Summerhayes (eds), *Understanding the Oceans: A Century of Ocean Exploration* (London and New York: UCL Press, 2001), pp. 27–48, p. 35.
8 'Report of the Submarine Telegraph Committee', Parliamentary Papers, 62, 1860, p. xxxv. Cited in Smith and Wise, *Energy and Empire*, p. 740.
9 Thomson, *Depths of the Sea*, p. 218.
10 Helen Rozwadowski, 'Depth Sounder', in Robert Bud and Deborah Jean Warner (eds), *Instruments of Science: An Historical Encyclopedia* (London: Science Museum, 1998), pp. 165–67.
11 Tizard et al., *Narrative of the Cruise*, vol. 1, Appendix C, p. 41.
12 Thomson, *Depths of the Sea*, p. 218.
13 Tizard et al., *Narrative of the Cruise*, vol. 1, p. 36.
14 Ibid.
15 Thomson, *Depths of the Sea*, p. 220.
16 Grace Fox, *Britain and Japan, 1858–1883* (Oxford: Clarendon Press, 1969), p. 270.
17 Rutherford Alcock, 'Address to the Royal Geographical Society', *Proceedings of the Royal Geographical Society of London* 21 (1876), p. 434.
18 United Kingdom Hydrographic Office, HMS *Sylvia* Sounding Book 1 and 2; Naval List 1870, National Archives, Kew; Anita McConnell, 'What's in a name? — C.W. Baillie', *History of Oceanography: Newsletter of the Commission of Oceanography, International Union of the History and Philosophy of Science*, 11, 1995, pp. 7–9, p. 7.
19 Thomson, *Depths of the Sea*, p. 220.
20 Hydrographic Department Minute Book 18, 7 May 1873, p. 130. National Archives, Kew, ADM 347.
21 Ibid., 4 June 1873, p. 161.
22 Ibid., 3–4 June 1873, pp. 160–61.
23 Ibid., 4 June 1873, pp. 160–61.
24 The expedition used a small collection of Baillie sounders during the voyage and their movements are similar enough to be referred to as 'the Baillie sounder'. Three of the sounders used during the *Challenger* Expedition from 1874 to 1876 are held at the Science Museum, London, and were viewed by the author.
25 Letter from Joseph Matkin to 'Tom' John Thomas Swann, 1 August 1873, in Rehbock (ed.), *At Sea with the Scientifics*, p. 95.
26 Letter from Matkin to Swann, 27 July 1873, in ibid., p. 94.
27 The sounding recorded was 2,275 fathoms (4,160 m) at Station 140. HD OD 268, HMS *Challenger* Sounding Book 1, Sounding Log, 24 August 1873, p. 66.
28 George Strong Nares, *Extracts from the Reports of Captain Nares to Rear Admiral Richards, Hydrographer of the Navy*, section 10.27.
29 Examining a Baillie rod in the collection of the Science Museum, the marks of the modifications made on board can still be seen today. Baillie sounder, Science Museum Object Number: 1876-825.
30 Nares, *Extracts from the Reports of Captain Nares to Rear Admiral Richards, Hydrographer of the Navy*, section 15.16.
31 Thomson, *The Voyage of the 'Challenger': The Atlantic*, vol. 1, p. 64.
32 Ibid.
33 Ibid.
34 Murray and Hjort, *The Depths of the Ocean*, p. 24.
35 Letter from Matkin to Swann, 7 April 1873, in Rehbock (ed.), *At Sea with the Scientifics*, p. 64.
36 Spry, *The Cruise of Her Majesty's Ship 'Challenger'*, p. 26.
37 Letter from Matkin to Swann, 5 June 1873, in Rehbock (ed.), *At Sea with the Scientifics*, p. 78.
38 Spry, *The Cruise of Her Majesty's Ship 'Challenger'*, p. 26.
39 Maury, *Physical Geography of the Sea*, p. 203.
40 Thomson, *Depths of the Sea*, pp. 207, 227.
41 Tizard et al., *Narrative of the Cruise*, vol. 1, pp. 65–66.

42 For a history of the Mariana Trench, see Albert Theberge, 'Thirty years of discovering the Mariana Trench', *Hydro International*, 2009, https://www.hydro-international.com/content/article/thirty-years-of-discovering-the-mariana-trench (accessed 10 April 2019).
43 Thomson, *Voyage of the 'Challenger': The Atlantic*, vol. 1, p. xvi.
44 For an analysis of *Challenger*'s sounding route see Erika Lynn Jones, 'Making the Ocean Visible: Science and Mobility on the *Challenger* Expedition, 1872–1895', PhD thesis, University College London, 2019, pp. 114–20 (available open access, UCL Discovery, https://discovery.ucl.ac.uk/id/eprint/10087545/).
45 Tizard et al., *Narrative of the Cruise*, vol. 1, pt. 1, Appendix B, 'Hydrographic instructions to Captain G.S. Nares, HMS *Challenger*', p. 35.
46 Ibid., pp. 35–36, p. 35.
47 Analysis based on comparisons of *Challenger*'s sounding charts and flows of global trade in the last quarter of the nineteenth century.
48 Spry, *The Cruise of Her Majesty's Ship 'Challenger'*, pp. 316–19.
49 Map of showing the telegraph lines in operation, under contract, and contemplated to complete the circuit of the globe (New York: J.H. Colton & Co., 1871), Library of Congress, https://www.loc.gov/item/86692708.
50 Tizard et al., *Narrative of the Cruise*, vol. 1, pt. 1, Appendix B, 'Hydrographic instructions to Captain G.S. Nares, HMS *Challenger*', p. 38.
51 Ibid.
52 Calculated from sounding figures given in Spry, *The Cruise of Her Majesty's Ship 'Challenger'*, pp. 316–18.
53 Alexander Agassiz, *A Contribution to American Thalassography: Three Cruises of the United States Coast and Geodetic Survey Steamer 'Blake', in the Gulf of Mexico, in the Caribbean Sea, and along the Atlantic Coast of t he United States, from 1877 to 1880*, 2 vols (Boston: Houghton, Mifflin and Company, 1888), vol. 1, p. 102.
54 Henry Cummings, *A Synopsis of the Cruise of the U.S.S. 'Tuscarora' from the Date of Her Commission to Her Arrival in San Francisco, Cal., Sept. 2nd, 1874* (San Francisco: Cosmopolitan Steam Printing Company, 1874), p. 50; Max Quanchi and John Robson, *Historical Dictionary of the Discovery and Exploration of the Pacific Islands* (Lanham, MD: Scarecrow Press, 2005).
55 Letter from Matkin to his mother, March 1873, in Rehbock (ed.), *At Sea with the Scientifics*, p. 55.
56 Deacon, *Scientists and the Sea*, p. 338; quote from Thomson, *Voyage of the 'Challenger': The Atlantic*, vol. 1, p. 177.
57 William Benjamin Carpenter, 'On the temperature of the deep-sea bottom and the conditions by which it is determined', *Proceedings of the Royal Geographical Society* 21 (1877), p. 297.
58 Rice, 'The *Challenger* Expedition', pp. 36–42; Tizard et al., *Narrative of the Cruise*, vol. 1, pt. 1, p. 97.
59 For Carpenter's debates on ocean circulation and *Challenger*'s temperature data, see Deacon, *Scientists and the Sea*, pp. 340–57.
60 National Maritime Museum, Greenwich, London, Reports to the Hydrographer by T.H. Tizard on HMS *Challenger* (1 March 1871 to 31 December 1874), TIZ/15.
61 See Dean Roemmich, W. John Gould and John Gilson, '135 years of global ocean warming between the *Challenger* Expedition and the Argo Programme', *Nature Climate Change*, 2 (June 2012), pp. 425–28.

CHAPTER 4

1 Herbert Swire, *The Voyage of the Challenger* (London: Golden Cockerel Press, 1938), p. 9.
2 P.F. Lingwood, 'The dispersal of the collections of H.M.S. *Challenger*: an example of the importance of historical research in tracing a systematically important collection', *History in the Service of Systematics: Papers From the Conference to Celebrate the Centenary of the British Museum (Natural History), 13–16 April 1981* 1 (1981), p. 71.
3 The object of this chapter is a mussel, *Cardita astartoides*, collected by the *Challenger* Expedition and described in the report by Edgar Albert Smith, *Report on the Lamellibranchiata Collected by H.M.S. Challenger During the Years 1873–76* [*Challenger Report*] (Edinburgh: HMSO, 1885), pp. 212–13. Now in the collection of the Natural History Museum: 1887.2.9.2868-70 (five dry specimens) and 1887.2.9.2870a-e (five wet specimens). The modern accepted species name is *Cyclocardia astartoides*.
4 Tizard et al., *Narrative of the Cruise*, vol. 1, pt. 1, p. 27–31.
5 Augustus A. Gould, *Mollusca & Shells, United States Exploring Expedition During the Years 1838, 1839, 1840, 1841, 1842*, vol. 12 (Boston: Gould & Lincoln, 1852), pp. vi–vii.
6 Ibid., p. xii.
7 Ibid., pp. ix–x.
8 After returning to Germany, the collection of molluscs was studied by Eduard von Martens, curator of the malacological section of the Museum für Naturkunde, Berlin. During his analysis, he discovered a new species that he named *Cardita astartoides*.
9 Tizard et al., *Narrative of the Cruise*, vol. 1, pt. 2, Appendix IV, p. 1027.
10 Ibid., p. 23.
11 Ibid., p. 36.
12 John Murray, 'On the Deep and Shallow-Water Marine Fauna of the Kerguelen Region of the Great Southern Ocean', *Transactions of the Royal Society of Edinburgh* 38 (1896), pp. 343–500, p. 348.
13 Tizard et al., *Narrative of the Cruise*, vol. 1, pt. 1, p. 330.
14 Natural History Museum (NHM) Earth Sciences Library, John James Wild, *H.M.S. Challenger: Diary of J.J. Wild* (1872–1876), 5 January 1872.
15 NHM Earth Sciences Library, Charles Wyville Thomson, *H.M.S. Challenger: Diary of Sir Wyville Thomson*, 1872–1876, 5 January 1842.
16 Lord George Campbell, *Log-Letters from the 'Challenger'* (London: Macmillan, 1877), p. 92.
17 Tizard et al., *Narrative of the Cruise*, vol. 1, pt. 2, p. 1027.
18 Letter from Joseph Matkin to 'Tom' John Thomas Swann, 1 February 1874, Kerguelen Island, in Rehbock (ed.), *At Sea with the Scientifics*, p. 133.
19 Richard Corfield, *The Silent Landscape: The Scientific Voyage of HMS Challenger* (Washington, DC: Joseph Henry Press, 2003), p. 152.
20 Campbell, *Log-Letters from the 'Challenger'*, p. 100.
21 Royal Geographical Society (RGS) archives, London, Pelham Aldrich's journal on HMS *Challenger*, Hopeful Harbor, 27 January 1874, p. 52, PEA/2.
22 Tizard et al., *Narrative of the Cruise*, vol. 1, pt. 1, p. 36.
23 Spry, *The Cruise of Her Majesty's Ship 'Challenger'*, p. 127.
24 Swire, *The Voyage of the Challenger*, p. 140.
25 Rozwadowski, *Fathoming the Ocean*, p. 137.
26 Gould, *Mollusca & Shells*, p. vii.

27 Spry, *The Cruise of Her Majesty's Ship 'Challenger'*, p. 388.
28 Ibid., p. 95.
29 Charles Wyville Thomson, *General Introduction to the Zoological Series of Reports* [*Challenger Report: Zoology*] (Edinburgh: HMSO, 1880), vol. 1, p. 24.
30 RGS archives, London, Aldrich, Hopeful Harbor, 27 January 1874, p. 49, PEA/2.
31 Smith, *Report on the Lamellibranchiata*, p. 3.
32 Pembre was also identified in an album as 'William Peniber'; see Eileen V. Brunton, *The Challenger Expedition, 1872–1876: A Visual Index* (London: Natural History Museum, 1994), p. 131; see a discussion of Pembre by Joseph Matkin in Rehbock (ed.), *At Sea with the Scientifics*, p. 128; Pembre died from what the ship's records describe as 'decline' in 'Abstract of Voyage of H.M.S. *Challenger*', in ibid., p. 371.
33 'Fluid Preserved Specimens', *American Museum of Natural History*, http://www.amnh.org/our-research/natural-science-collections-conservation/general-conservation/health-safety/fluid-preserved-specimens (accessed 10 April 2019).
34 Thomson, *General Introduction to the Zoological Series of Reports*, vol. 1, p. 23.
35 Ibid., p. 24.
36 NHM archives, London, HMS *Challenger* contents of packing cases made on board, 44NHM ALMA.
37 Thomson, *General Introduction to the Zoological Series of Reports*, vol. 1, p. 23.
38 NHM archives, contents of packing cases made on board, 44NHM ALMA. Listed as 'Shells from dredge 17 January 1874, 20 to 60 fathoms, Kerguelen', Box A. 22.
39 NHM, HMS *Challenger* contents of packing cases made on board, 44NHM ALMA.
40 Thomson, *General Introduction to the Zoological Series of Reports*, vol. 1, p. 24.
41 Tizard et al., *Narrative of the Cruise*, vol. 1, pt. 1, pp. 7–8.
42 Thomson, *General Introduction to the Zoological Series of Reports*, vol. 1, p. 23.
43 Nares, *Extracts from the Reports*, section 15.18. See also Tizard et al., *Narrative of the Cruise*, vol. 1, pt. 1, p. 3.
44 Tizard et al., *Narrative of the Cruise*, vol. 1, pt. 1, p. 4.
45 For *Challenger* in Australia, see David F. Branagan, 'The *Challenger* Expedition and Australian science', *Proceedings of the Royal Society of Edinburgh Section B: Biological Sciences* 73 (1972), pp. 85–95.
46 RGS archives, London, Aldrich, 27 January 1874, p. 66, PEA/2.
47 Spry, *The Cruise of Her Majesty's Ship 'Challenger'*, p. 157.
48 William Maddock, *Visitors' Guide to Sydney* (Sydney: William Maddock, 1872), p. 5.
49 John Jeremy, *Cockatoo Island: Sydney's Historic Dockyard* (Sydney: University of New South Wales, 2005), p. 9.
50 Ibid., pp. 9–12.
51 For relevant history of the Royal Mail in the nineteenth century, see William Lewins, *Her Majesty's Mails: A History of the Post Office and an Industrial Account of its Present Condition* (London: Sampson Low, Son, and Marston, 1865), pp. 231–37; Martin J. Daunton, *Royal Mail: The Post Office Since 1840* (London: Bloomsbury, 2015), pp. 146–90.
52 NHM archives, 44NHM ALMA, 'Catalogue of Boxes and Casks and their Contents sent from H.M.S. *Challenger* at Sydney to England, 8 June 1874'.
53 Peter Plowman, *Across the Pacific: Liners from ANZ to North America* (Kenthurst: Rosenberg Publishing, 2010), pp. 29, 31–34, 36–38.
54 Maddock, *Visitors' Guide to Sydney*, p. iv.
55 Letter of 22 August 1871 to Members of the Legislature of Australia, in Henry Parkes, *Fifty Years in the Making of Australian History*, vol. 1 (1892) (New York: Books for Librarians Press, 1971).
56 Letter from Matkin to his mother, Sydney, n/d, in Rehbock (ed.), *At Sea with the Scientifics*, p. 163.
57 Plowman, *Across the Pacific*, p. 37.
58 Ibid.
59 Bruce C. Cooper, 'Postal history of the first transcontinental railroad', 2014, http://cprr.org/Museum/Ephemera/Postal.html (accessed 10 April 2019).
60 Personal communication with Pamela Henson, Historian of the Smithsonian Institution.
61 Stephen E. Ambrose, *Nothing Like It in the World: The Men Who Built the Transcontinental Railroad 1863–1869* (New York: Touchstone, 2000), pp. 19–22.
62 Thomson, *General Introduction to the Zoological Series of Reports*, vol. 1, p. 26.
63 Smith, *Report on the Lamellibranchiata*, Plate XV, figs 2–2c.
64 Letter from John Murray to Alfred Russel Wallace, 20 May 1889, in *Wallace Letters Online*, http://www.nhm.ac.uk/resources/research-curation/projects/wallace-correspondence/transcripts/pdf/WCP2407_L2297.pdf (accessed 1 March 2016). See also Alfred Russel Wallace, *Darwinism: An Exposition on the Theory of Natural Selection with Some of Its Applications* (London and New York: Macmillan, 1889).
65 Murray, 'On the Deep and Shallow-Water Marine Fauna of the Kerguelen Region', pp. 343–44.

CHAPTER 5

1 Thomson, *The Voyage of the 'Challenger'*, vol. 1, p. 46.
2 Helmut Gernsheim, Naomi Rosenblum, Andy Grundberg and Beaumont Newhall, 'History of photography', *Encyclopedia Britannica*, November 2021, https://www.britannica.com/technology/photography (accessed 10 July 2022).
3 James R. Ryan, *Photography and Exploration* (London: Reaktion Books, 2013), p. 35.
4 See Brunton, *The Challenger Expedition*.
5 Photography on board *Challenger* served many purposes, see Stephanie Hood, 'Science, objectivity and photography in the nineteenth century: photographs from the voyage of HMS *Challenger* 1872–1876', MSc thesis, University College London, 2013.
6 Brunton, *The Challenger Expedition*, p. 15; Hood, 'Science, objectivity and photography in the nineteenth century', p. 34.
7 Brunton identified six *Challenger* albums belonging to Henry Moseley, Rudolf von Willemoes-Suhm, John Young Buchanan, Thomas Henry Tizard, John Hynes and Robert Higham. These albums alone required the manufacture of some 1,785 prints.
8 The photographic albums of Assistant Paymaster John Hynes are held in the collection of the National Maritime Museum, Greenwich. ALB0174 (139 photographs relating to the first part of the voyage in 1873), ALB0175 (143 photographs relating to the mid-part of the voyage, 1873–75) and ALB0176 (89 photographs relating to the latter part of the voyage, 1875–76). The NMM also holds the albums of Navigating Lieutenant Thomas Henry Tizard (ALB0859).
9 Jones, 'Making the Ocean Visible: Science', PhD thesis, University College London, 2019, pp. 191–243 (available open access, UCL Discovery,

https://discovery.ucl.ac.uk/id/eprint/10087545/).

10 In 1869 the Royal Engineers Establishment was renamed the School of Military Engineering. John Falconer, 'Royal Engineers (RE)', in *The Oxford Companion to the Photograph*, Robin Lenman and Angela Nicholson (eds) (Oxford: Oxford University Press, 2005), pp. 545–46, p. 545.

11 James R. Ryan, *Picturing Empire: Photography and the Visualization of the British Empire* (London: Reaktion Books, 1997), p. 78.

12 Frederic A. Sharf, *Abyssinia, 1867–1868: Artists on Campaign, Watercolors and Drawings from the British Expedition under Sir Robert Napier* (Hollywood, CA: TSEHAI Publishers, 2003), p. 96.

13 Ibid.

14 William de Wiveleslie Abney, 'Photography in the Arctic seas', *British Journal of Photography* 20 (1873), p. 57.

15 Emphasis is in the original. William de Wiveleslie Abney, *Instruction in Photography: For Use at the S.M.E., Chatham* (Chatham: School of Military Engineering, 1871), p. 1.

16 Major General Whitworth Porter, *History of the Corps of Royal Engineers*, vol. 2 (Chatham: The Institution of Royal Engineers, 1889), p. 188.

17 Abney, 'Photography in the Arctic seas', p. 57.

18 Peter D. Hingley, 'The first photographic eclipse?', *Astronomy & Geophysics* 42 (2001), pp. 118–22, p. 120.

19 William de Wiveleslie Abney, 'Photographic operations in Egypt in connection with the late transit of Venus', *British Journal of Photography* 22 (1875), pp. 153–54, p. 153.

20 Ryan, *Photography and Exploration*, p. 10. Other examples of expeditions pushing the boundaries of photography soon after its invention and in the NMM collection include William Domville's calotype negatives and prints from the 1852 Franklin search expedition, and Edward Inglefield's wet-plate negatives from the 1854 search expedition.

21 For example, the back pages of Abney's book *Emulsion Processes in Photography* contained adverts for a wide range of photographic companies and apparatus for sale. William de Wiveleslie Abney, *Emulsion Processes in Photography* (London: Piper & Carter, 1878), p. 224.

22 Ibid.

23 For Abney's recommended photographic apparatus, see ibid.

24 Brunton, *The Challenger Expedition*, p. 17.

25 Ibid., p. 25. Ship Photo No. 1. This is the first recorded photograph; others may have existed but were destroyed.

26 Abney, *Instruction in Photography*, p. 1. See 'Henry Peach Robinson', July 2022, *Encyclopedia Britannica*, https://www.britannica.com/biography/Henry-Peach-Robinson#ref12958 (accessed 20 July 2022).

27 Henry Peach Robinson, *Pictorial Effect in Photography* (London: Piper & Carter, 1879), p. 16.

28 For example, see HMS *Challenger* Meteorological Table for the Month of June 1873, TIZ/63/2.

29 Thomson, *The Voyage of the 'Challenger'*, vol. 1, p. 46.

30 Ibid.

31 Such as the image of Bahia, Brazil, ALB0174.107.

32 For example, for his book, *At Anchor: A Narrative of Experiences Afloat and Ashore During the Voyage of H.M.S. 'Challenger' from 1872 to 1876* (London and Belfast: M. Ward and Co., 1878).

33 Newbold's known photographs roughly align with Brunton's photographs 1–276. Of these, Newbold took 27 photographs that had people as their main subject, compared with 248 photographs of various landscapes, harbours, anchorages, plants and buildings.

34 Brunton, *The Challenger Expedition*, p. 20.

35 Henry Nottidge Moseley, *Notes by a Naturalist: Observations Made During the Voyage of H.M.S. 'Challenger' Round the World in the Years 1872–1876* (London: John Murray, 1892), p. 24.

36 Tizard et al., *Narrative of the Cruise*, vol. 1, pt. 1, p. 143; William Botting Hemsley, *Report on the Botany of the Bermudas and Various Other Islands of the Atlantic and Southern Oceans* [*Challenger Report: Botany*], vol. 1., pt. 2 (Edinburgh: HMSO, 1885).

37 Lady Lefroy created an album of 83 paintings of plants found in Bermuda, made during the period 1871–77 when she and her husband resided on the islands. It is held by the Bermuda National Trust, https://bnt.bm/heritage/collections/the-lady-lefroy-collection/ (accessed 11 July 2022).

38 For a discussion of female botanical collectors and illustrators working in colonial settings in the nineteenth century, see Hannah Wills, Sadie Harrison, Erika Lynn Jones, Farrah Lawrence-Mackey and Rebecca Martin (eds), *Women in the History of Science: A Sourcebook* (London: UCL, 2023).

39 Moseley, *Notes by a Naturalist*, p. 16.

40 Ibid., pp. 15–25.

41 Thomson, *The Voyage of the 'Challenger'*, vol. 2, pp. 100–01.

42 'The Challenger at St. Paul's Rocks', *Illustrated London News*, 1 November 1873, p. 414.

43 Thomson, *The Voyage of the 'Challenger'*, vol. 2, pp. 102–08.

44 Moseley, *Notes by a Naturalist*, p. 58.

45 Tizard et al., *Narrative of the Cruise*, vol. 1, pt. 1, Plate IV.

46 A photograph of Wild's drawing appears in Hynes's album (ALB0174.133).

47 *Illustrated London News*, vol. LXIII, 1 November 1873, p. 1.

48 Spry, *The Cruise of Her Majesty's Ship 'Challenger'*, p. 60.

49 Letter from Joseph Matkin to Sarah Craxford Matkin, Prince Edward Island, Christmas Day 1873, in Rehbock (ed.), *At Sea with the Scientifics*, p. 130.

50 Brunton, *The Challenger Expedition*, pp. 16–17.

51 Anne Tove Austbo, 'Picturing seafarers: new perspectives on the photographic crew portrait ca. 1870–1940', paper given at *Old and New Uses of the Oceans, 8th IMHA International Congress of Maritime History*, 29 June 2022, University of Porto, Portugal.

52 Robinson, *Pictorial Effect in Photography*, p. 49.

53 For an analysis of *Challenger*'s images of icebergs, see Rosamunde Codling, 'HMS *Challenger* in the Antarctic: Pictures and Photographs from 1874', *Landscape Research*, 22, 1997, pp. 191–208.

54 Tizard et al., *Narrative of the Cruise*, vol. 1, pt. 1, pp. 397–98; Eric Linklater, *The Voyage of the Challenger* (London: John Murray, 1972), p. 85.

55 Codling, 'HMS *Challenger* in the Antarctic', pp. 195–97.

56 NMM, Journal of John Hynes, HMS *Challenger*, December 1873–March 1874, JOD/15/1, p. 108.

57 Tizard et al., *Narrative of the Cruise*, vol. 1, pt. 1, p. 430.

58 Ibid., vol. 1, pt. 1, p. 433.

59 Codling, 'HMS *Challenger* in the Antarctic', pp. 195–97.

60 NMM, ALB0175.14, ALB0175.20, ALB0175.22, ALB0175.24, ALB0175.26

and ALB0175.27. Eight photographs of icebergs were published in the *Challenger Report*. See Tizard et al., *Narrative of the Cruise*, vol. 1, pt. 1, Plates X–XIII

61 Jones, 'Making the Ocean Visible', pp. 229–32.

62 See Julie F. Codell, 'Victorian portraits: re-tailoring identities', *Nineteenth-Century Contexts: An Interdisciplinary Journal* 34:4 (2012), pp. 493–516.

63 From Tizard et al., *Narrative of the Cruise*, vol. 1, pt. 1, pp. 478, 480, and Moseley, *Notes by a Naturalist*, p. 247, a four-man gig crew plus a coxswain were photographed together on board *Challenger* at Tongatapu as a group. This man was photographed not with the boat crew (ALB0175.47) but alone during the same period that the pilot was on board. While not confirmed, the author considers his identity as pilot to be plausible.

64 Austbo, 'Picturing seafarers'.

65 For an example of ethnographic descriptions recorded on the *Hassler* Expedition, see Erika Jones, 'Elizabeth Cabot Agassiz (1822–1907): Voyage of the *Hassler* (1871–2)', in Wills, Harrison, Jones, Lawrence-Mackey and Martin (eds), *Women in the History of Science*.

66 Moseley, *Notes by a Naturalist*, p. ix.

67 The widespread acceptance of polygenesis was forged through influential books such as Robert Knox's *The Races of Men* (1850) and prestigious scientists such as Louis Agassiz.

68 Moseley, *Notes by a Naturalist*, pp. ix–x.

69 The British Museum holds some 176 ethnographic objects related to the *Challenger* Expedition. See https://www.britishmuseum.org/collection/term/BIOG244795 (accessed 11 July 2022).

70 Moseley, *Notes by a Naturalist*, p. 247.

71 On *Challenger*'s return, Moseley gave the human remains to Sir William Turner at the University of Edinburgh. Turner's resulting study added to scientific debates on human origins and promoted ideas that supported the early twentieth-century eugenics movement. As far as the author is aware, the human remains collected by *Challenger* have not been returned to any of the communities. See William Turner, *Report on the Human Crania and Other Bones of the Skeletons Collected by H.M.S. Challenger During the Years 1873–1876* [*Challenger Report: Zoology*] (Edinburgh: HMSO, 1884, 1886), vol. 10, pt. 29 and vol. 16, pt. 47.

72 See Moseley, *Notes by a Naturalist*, p. 247.

73 Moseley, Notes by a Naturalist, p. 247.

74 Tizard et al., *Narrative of the Cruise*, vol. 1, p. 507. The photograph appears as Plate XIX.

75 Lord George Campbell, *Log-Letters from the 'Challenger'* (London: Macmillan, 1877), p. 102.

76 Letter from Joseph Matkin to his mother, 19 December 1874, Hong Kong, in Rehbock (ed.), *At Sea with the Scientifics*, p. 211.

77 HMS *Victor Emmanuel* (launched as *Repulse*, 1855), see the Victorian Royal Navy, HMS *Victor Emmanuel* (pdavis.nl) (accessed 13 July 2022).

78 National Archives, Kew, ADM 188/24/51552.

79 Note that the author's observation is for Hynes's album only.

80 'Wet plate photography — troubleshooting', https://yvettebessels.com/?page_id=3610 (accessed 25 January 2017).

81 Hood, 'Science, objectivity and photography in the nineteenth century', p. 14.

82 Photography on board *Challenger* was a changing practice that was shaped by the voyage and the ocean environment. Jones, 'Making the Ocean Visible', pp. 191–243.

83 Ibid.

84 Spry, *The Cruise of Her Majesty's Ship 'Challenger'*, p. 204.

85 See Wild, *At Anchor*, p. 129.

86 Rudolf von Willemoes-Suhm quoted in F.E. Schulze, *Report on the Hexactinellida Collected by H.M.S. Challenger* [*Challenger Report: Zoology*] (Edinburgh: HMSO, 1887), vol. 21, pt. 53, p. 61. Also see *Wild, At Anchor*, pp. 129–30.

87 Wild, *At Anchor*, p. 129.

88 Tizard et al., *Narrative of the Cruise*, vol. 1, pt. 2, p. 651; also see Moseley, *Notes by a Naturalist*, pp. 351–52.

89 Spry, *The Cruise of Her Majesty's Ship 'Challenger'*, p. 253.

90 See Brunton, *The Challenger Expedition*, pp. 154–55, 162–63, 178. For Royal Society instructions to *Challenger* naturalists regarding the ethnographic study, see Tizard et al., *Narrative of the Cruise*, vol. 1, p. 32.

91 NMM, 'Death of Paymaster-in-Chief at Portsmouth', newspaper clipping dated 26 March 1926 in the journal of John Hynes, JOD/15/1.

92 For the development of the album, see Anna Dahlgren, 'The ABC of the modern photo album', in Jonathan Carson, Rosie Miller and Theresa Wilkie (eds), *The Photograph and the Album: Histories, Practices, Futures* (Edinburgh: MuseumsEtc, 2013), pp. 77–103.

93 NMM, JOD/15/1.

94 Tizard et al., *Narrative of the Cruise*, vol. 1, p. 761.

95 While Brunton's index has begun the cataloguing process, identifying information is generally limited to the original titles written in the albums.

96 For example, see Emma Zuroski, 'Situating the local in a global expedition: HMS *Challenger* Expedition in New Zealand, 1874', *Journal of the Royal Society of New Zealand* 47:1 (2017), pp. 107–11.

97 Research in this area is ongoing, for instance Erika Jones and Rebecca Martin are currently studying and cataloguing *Challenger* photographs at the National Maritime Museum, Greenwich. New work has brought attention to Charlie Collins, *Challenger*'s head stoker: see Pearson, *A Challenger's Song*.

CHAPTER 6

1 George Russell Agassiz (ed.), *Letters and Recollections of Alexander Agassiz: with a sketch of his life and work* (Boston: Houghton, Mifflin and Company, 1913), pp. 159–60.

2 Spry, *The Cruise of Her Majesty's Ship 'Challenger'*, p. 314.

3 The specimen this chapter follows, *Salenocidaris varispina* Agassiz 1869, is held in the collection of the National Museum of Natural History, Washington, DC, USNM Catalogue Number 17543. Preparation: Alcohol (Ethanol). Accession Number: 023665. For a full record, see http://n2t.net/ark:/65665/3608d7c33-ac65-40e9-9540-02b1b9915dd6.

4 Depending on how they are counted, the *Zoology* section of the *Challenger Report* consists of 32 or 40 volumes. John Murray claimed the larger number. See Martyn Low and Neal Evenhuis, 'Dates of publication of the Zoology parts of the *Report on the Scientific Results of the Voyage of H.M.S. Challenger During the Years 1873–76*', *Zootaxa* 3701 (2013), pp. 401–20.

5 Although outside the scope of this chapter, in addition to his zoological work, Louis Agassiz was a vocal proponent of polygenism, a theory that sought to rank and segregate people in different races. See Saima S. Iqbal, 'Louis Agassiz, under a microscope', *The Harvard Crimson*, 18 March 2021, https://www.thecrimson.com/article/2021/3/18/louis-agassiz-scrut/ (accessed 9 August 2021).
6 Agassiz (ed.), *Letters and Recollections of Alexander Agassiz*, pp. 25–29, quote p. 29.
7 Ibid., p. 53.
8 Ibid., pp. 96–98.
9 Ibid., pp. 91, 93.
10 A note on usage of taxonomic names: 'Echinoidea' is the name of the class and a plural noun for all members of the class. Modern scientists refer to sea urchins as 'echinoids', while Agassiz and his contemporaries used the term 'echini'. See 'Echinoids', British Geological Survey, https://www.bgs.ac.uk/discovering-geology/fossils-and-geological-time/echinoids/ (accessed 2 August 2022).
11 Letter from Alexander Agassiz to Charles Darwin, 9 December 1872, in Agassiz (ed.), *Letters and Recollections of Alexander Agassiz*, pp. 119–20.
12 John Murray, 'Alexander Agassiz: his life and scientific work', *Science* 54 (1911), pp. 873–87, p. 875.
13 Ibid.
14 George Russell Agassiz (ed.), *Letters and Recollections of Alexander Agassiz*, pp. 97, 98.
15 Sir Charles Wyville Thomson, University of Edinburgh, https://www.ed.ac.uk/about/people/plaques/thomson (accessed 2 August 2022).
16 Murray, 'Alexander Agassiz: his life and scientific work', p. 876.
17 Agassiz (ed.), *Letters and Recollections of Alexander Agassiz*, p. viii.
18 Thomson distributed the materials of the 1868, 1869 and 1870 cruises to experts and 15 zoological reports appeared in British scientific journals from 1871 to 1885.
19 Letter from Alexander Agassiz to Charles Darwin, 22 October 1870, in Agassiz (ed.), *Letters and Recollections of Alexander Agassiz*, p. 113.
20 See Thomson, *Depths of the Sea*, pp. 189–96.
21 Edinburgh University Library (EUL), Letter from Alexander Agassiz to Wyville Thomson, 28 May 1877, Coll-96.
22 Letter from Alexander Agassiz to Elizabeth Agassiz, 30 October 1870, in Agassiz (ed.), *Letters and Recollections of Alexander Agassiz*, p. 113.
23 Ibid.
24 Agassiz (ed.), *Letters and Recollections of Alexander Agassiz*, p. 122.
25 Alexander Agassiz, *Revision of the Echini, Illustrated Catalogue of the Museum of Comparative Zoölogy at Harvard College*, No. 7 (Cambridge, MA: Cambridge University Press, 1872).
26 P.F. Lingwood, 'The dispersal of the collections of H.M.S. *Challenger*: an example of the importance of historical research in tracing a systematically important collection', *History in the Service of Systematics: Papers from the Conference to Celebrate the Centenary of the British Museum (Natural History), 13–16 April 1981* 1 (1981), pp. 71–77, p. 71.
27 Thomson, *General Introduction to the Zoological Series of Reports*, vol. 1, p. 27.
28 Ibid.
29 Ibid.
30 Deacon, *Scientists and the Sea*, pp. 369–70.
31 Agassiz (ed.), *Letters and Recollections of Alexander Agassiz*, p. 157.
32 Ibid., p. 124.
33 Ibid.
34 Ibid., p. 125.
35 Murray, 'Alexander Agassiz: his life and scientific work', p. 879.
36 Deacon, *Scientists and the Sea*, p. 367.
37 Peter Martin Duncan, 'Zoology of the "Challenger" Expedition', *American Magazine of Natural History* 19 (1877), pp. 506–08, p. 508.
38 'Copy of a Memorandum delivered to Professor Owen from Albert Günther of the British Museum', 6 June 1876, in *Correspondence Concerning the Specimens and Collections Made by the 'Challenger' Expedition* (London: Taylor and Francis, 1877), p. 234.
39 Deacon, *Scientists and the Sea*, p. 368.
40 Museum of Comparative Zoology (MCZ), Harvard University, Cambridge, Massachusetts, Letter from John Murray to Alexander Agassiz, 4 July 1877, bAg654.10.1.
41 MCZ, Letter from John Murray to Alexander Agassiz, 13 April 1877, bAg654.10.1.
42 James D. Dana, B. Silliman, and E. S. Dana, 'Sir Wyville Thomson, and the working up of the "Challenger" collections', *The American Journal of Science and Arts* 14 (1877), pp. 161–63, p. 162.
43 MCZ, *Annual Report of the Curator of the Museum of Comparative Zoology at Harvard College, to the President and Fellows of Harvard College for 1877–1878* (Cambridge: John Wilson and Son, 1878), p. 10.
44 Agassiz, *Revision of the Echini*, p. xi.
45 *Annual Report of the Curator of the Museum of Comparative Zoology*, p. 7.
46 Alexander Agassiz, 'Preliminary report on the Echini and star fishes, dredged in deep water between Cuba and the Florida Reef by L.F. De Pourtales', *Bulletin of the Museum of Comparative Zoology at Harvard College* 1 (1869), pp. 255–56 p. 256; MCZ object holotype of *Salenocidaris varispina* A. Agassiz, 1869. Whole animal (dry), collected off Double Head Shot Key, Florida, USA, 315 fathoms. Invertebrate Zoology ECH-385.
47 Alexander Agassiz, *Report on the Echinoidea Dredged by H.M.S. Challenger, During the Years 1873–1876* [*Challenger Report: Zoology*] (Edinburgh: HMSO, 1881), vol. 3, pt. 9, p. 55.
48 Ibid., p. 1.
49 See Agassiz, *A Contribution to American Thalassography*. Charles Sigsbee was in command of the USS *Blake* from 1875 to 1878.
50 Agassiz, *Report on the Echinoidea*, pp. 1, 55, 90.
51 Alexander Agassiz, 'Paleontological and Embryological development', *Proceedings of the American Association for the Advancement of Science* 29 (1880), pp. 389–414; Mary P. Winsor, *Reading the Shape of Nature: Comparative Zoology at the Agassiz Museum* (Chicago: University of Chicago Press, 1991), pp. 154–56.
52 Agassiz, 'Paleontological and Embryological Development', p. 413.
53 Ibid.
54 Winsor, *Reading the Shape of Nature*, p. 156.
55 Agassiz, 'Paleontological and Embryological Development', p. 413.
56 Theodore Lyman, *Report on the Ophiuroidea Dredged by H.M.S. Challenger During the Years 1873–1876* [*Challenger Report: Zoology*] (Edinburgh: HMSO, 1882), vol. 5, pt.14, p. 1.
57 Yoshihiro Tanimura and Yoshiaki Aita (eds), 'Haeckel Radiolaria Collection and the HMS *Challenger* Plankton Collection', National Museum of Nature and Science Monographs, No. 40, Tokyo, 2009, pp. 35–45; Ernst Haeckel, *Report on the Radiolaria Collected by*

H.M.S. Challenger During the Years 1873–1876 [*Challenger Report: Zoology*] (Edinburgh: HMSO, 1887), vol. 18, pt. 40.

58 While Haeckel's work on the *Challenger* Radiolaria is not contentious, it is important to recognise that he used his scientific platform to promote damaging theories about the hierarchy of races, including eugenics. See John P. Jackson and Nadine M. Weidman (eds), *Race, Racism, and Science: Social Impact and Interaction* (New Brunswick: Rutgers University Press, 2005), p. 87. David Lazarus, 'A brief review of radiolarian research', *Paläntologische Zeitschrift* 79:1 (31 March 2005), pp. 183–200.

59 Albert Günther, *Report on the Deep-Sea Fishes Collected by H.M.S. Challenger During the Years 1873–1876*, [*Challenger Report: Zoology*] (Edinburgh: HMSO, 1887), vol. 22, pt. 57, p. i.

60 Ibid., p. ii.

61 Georg Ossian Sars, *Report on the Schizopoda Collected by H.M.S. Challenger During the Years 1873–1876* [*Challenger Report: Zoology*] (Edinburgh: HMSO, 1885), vol. 13, pt. 37, pp. 1–2.

62 S.I. Smith, 'Recent Challenger Reports: 'Report on the Schizopoda' (vol. 13) by Prof. G. O. Sars. London, Government, 1885', *Science* 7:162 (12 March 1886), pp. 249–50, quotes p. 249; Sars, *Report on the Schizopoda*, vol. 13, pt. 37.

63 Thomas Henry Huxley and Paul Pelseneer, *Report on the Specimen of the Genus Spirula Collected by H.M.S. Challenger* [*Challenger Report: Zoology*] (Edinburgh: HMSO, 1895), Appendix, pt. 83, p. 3.

64 See Musée Océanographique, www.musee.oceano.org/en/ (accessed 9 August 2022).

65 Harold L. Burstyn, 'Science pays off: Sir John Murray and the Christmas Island phosphate industry, 1886–1914', *Social Studies of Science* 5:1 (February 1975), pp. 5–34.

66 Murray and Hjort, *The Depths of the Ocean*.

67 MCZ, *Challenger* checklist Ophiur and Astropht, *c.*1881, bAu 1822.44.3.

68 At the time of this research, there were 51 *Challenger* Echinoidea specimens in the MCZ collection with accession number 443978. Excel spreadsheet created 24 June 2016 by MCZ curators, '*Challenger* specimens held at the MCZ'.

69 Smithsonian Institution Archives, Accession Records, Record Unit 305.

70 Smithsonian Institution Archives, Letter to Mr Goode from A. Günther, British Museum (Natural History), 8 September 1890, USNM Accession Record 23665.

CHAPTER 7

1 John Murray and Alphonse-François Renard, *Report on Deep-Sea Deposits Based on the Specimens Collected During the Voyage of H.M.S. Challenger in the Years 1872 to 1876* [*Challenger Report*] (Edinburgh: HMSO, 1891), Editorial Note, n.p.

2 Harold L. Burstyn, '"Big science" in Victorian Britain: the *Challenger* Expedition (1872–6) and its Report (1881–95)', in Margaret Deacon, Tony Rice and Colin Summerhayes (eds), *Understanding the Oceans: A Century of Ocean Exploration* (Boca Raton, FL: CRC, 2002), pp. 51–55, p. 51.

3 Deacon, *Scientists and the Sea*, p. 375.

4 Tizard et al., *Narrative of the Cruise*.

5 *Physics and Chemistry* [*Challenger Report*], 2 vols (Edinburgh: HMSO, 1884, 1889).

6 Aileen Fyfe notes that the change in publishing was not only technologically determined by steam power, but also involved the forces of the market, cultural and social factors. Aileen Fyfe, *Steam-Powered Knowledge: William Chambers and the Business of Publishing, 1820–1860* (Chicago: University of Chicago Press, 2012), pp. 2–3.

7 For relevant history of the Royal Mail in the nineteenth century, see William Lewins, *History of the Post Office and an Industrial Account of its Present Condition* (London: Sampson Low, Son, and Marston, 1865), pp. 231–37; Martin J. Daunton, *Royal Mail: The Post Office Since 1840* (London: Bloomsbury, 2015), pp. 146–90.

8 *Annual Report of the Board of Regents of the Smithsonian Institution* (Washington, DC: Smithsonian Institution, 1858), pp. 242–43.

9 Thomson, *The Voyage of the 'Challenger': The Atlantic*, vol. 1., p. xx.

10 Thomson, *General Introduction to the Zoological Series of Reports*, vol. 1, pt. 1, p. xi.

11 Moseley, *Notes by a Naturalist*, p. 517.

12 Tizard et al., *Narrative of the Cruise*, vol. 1, pt. 2, Appendix VI. Bibliography, giving the Titles of Books, Reports, and Memoirs, referring to the Results of the Challenger Expedition, pp. 1053–63, p. 1059.

13 Henry Moseley, 'Botanical notes in letters addressed to Sir J.D. Hooker', *Nature* 9 (1874), pp. 369, 388, 450, 485; and 10 (1874), p. 165.

14 See chapter 3 for details on how sediments were collected using the Baillie sounder.

15 John Murray, 'Preliminary reports to Professor Wyville Thomson, F.R.S., Director of the Civilian Scientific Staff, on Work Done on Board the "*Challenger*"', *Proceedings of the Royal Society of London* 24 (1875), pp. 471–544, p. 471.

16 Ibid., p. 471.

17 The individual preliminary reports to Thomson appear as 'Reports from the "*Challenger*"' in *Proceedings of the Royal Society of London* 24 (1875), pp. 463–636.

18 Charles Wyville Thomson, 'Progress of the "*Challenger*" Report', *Nature* 18:463 (1878), pp. 534–35, p. 534; Lynn Nyhart, 'Voyaging and the scientific expedition report, 1800–1940', in Rima D. Apple, Gregory J. Downey and Stephen L. Vaughn (eds), *Science in Print Essays on the History of Science and the Culture of Print* (Madison, WI: University of Wisconsin Press, 2012), pp. 65–86.

19 Thomson, *General Introduction to the Zoological Series*, vol. 1, p. 25.

20 Ibid.

21 Smithsonian Libraries, Nathaniel Philbrick, 'The United States Exploring Expedition, 1838–1842', January 2004, https://www.sil.si.edu/DigitalCollections/usexex/learn/Philbrick.htm (accessed 10 August 2022).

22 Thomson, *General Introduction to the Zoological Series*, pp. 25–28.

23 Ibid., p. 29.

24 Ibid., p. xiii.

25 Ibid., p. 28.

26 Thomson, *Depths of the Sea*, p. vii.

27 Thomson, 'Progress of the "*Challenger*" Report', pp. 534–35.

28 Review by G. A. L., *Geological Magazine* 4:4 (1877), pp. 174–76, p. 174.

29 John Murray, 'On the distribution of volcanic debris over the floor of the ocean, its character, source, and some of the products of its disintegration and decomposition', *Proceedings of the Royal Society of Edinburgh* 9 (1877), pp. 247–61, p. 258.

30 Ibid.
31 Archibald Geikie, 'Lehrbuch der Mineralien und Felsartenkunde', *Nature* 3 (1871), pp. 283–85, pp. 283–84. Geikie wrote about the state of British microscopic petrography in his review of this book by 'Dr F. Senft of Jena'.
32 Ferdinand Zirkel, *Microscopical Petrography*, vol. 6, in Clarence King, Geologist-in-Charge, *United States Report of the Geological Exploration of the Fortieth Parallel* (Washington, DC: Government Printing Office, 1876).
33 Ibid., p. 12.
34 Jean-Pierre Henriet, 'The Face of the Ocean: Alphonse-François Renard (1842–1903) and the Rise of Marine Geology', *Sartoniana* 23 (2010), pp. 47–80, pp. 55–58.
35 See chromolithographic plates in Charles de la Vallée Poussin and Alphonse-François Renard, *Mémoire sur les caractères minéralogiques et stratigraphiques des roches dites plutoniennes de la Belgique et de l'Ardenne française* (Brussels: Académie Royale de Belgique, 1876).
36 Henriet, 'The Face of the Ocean', p. 56.
37 Ibid., p. 59.
38 Murray and Renard, *Report on Deep-Sea Deposits*, p. 17.
39 Ibid., p. ix.
40 MCZ, Letter from John Murray to Alexander Agassiz, 1 March 1881, bAg 654.10.1.
41 Murray and Renard, *Report on Deep-Sea Deposits*, p. 27. The petrographers who worked on the materials are named only as 'Sipocz, Hornung and Klement'.
42 University of Edinburgh, Centre for Research Collections, HMS *Challenger* Papers, 1872–1876, Coll-46 at Gen. 28-31, note from Belgian printer G. Severeyns addressed to Renard, 25 April 1883.
43 Ferdinand Zirkel, *Die Mikroskopische Beschaffenheit Der Mineralien Und Gesteine* (Leipzig: Verlag von Engelmann, 1873), pp. 27–28.
44 Murray and Renard, *Report on Deep-Sea Deposits*. Plates I–XXIX are variously credited to lithographers George West, Dr J. Heitzmann, Renard and some are anonymous. They were printed in Edinburgh, Vienna and Brussels respectively.
45 University of Edinburgh, Centre for Research Collections, HMS *Challenger* Papers, Coll-46.
46 Murray and Renard, *Report on Deep-Sea Deposits*, Plate XXIII.
47 University of Edinburgh, Centre for Research Collections, Correspondence of Alexander Agassiz (1867–1910), Coll-96, letters to John Murray and Charles Wyville Thomson.
48 See obituary, 'Sir Charles Wyville Thomson' in *Nature* 25 (16 March 1882), pp. 467–68, p. 467; Deacon, *Scientists and the Sea*, p. 372.
49 Tizard et al., *Narrative of the Cruise*, vol. 1, pt. 1, p. viii.
50 University of Edinburgh, Centre for Research Collections, HMS *Challenger* Papers, Coll-46.
51 Ibid.
52 Murray and Renard, *Report on Deep-Sea Deposits*, p. xxviii.
53 See Michael S. Reidy and Helen M. Rozwadowski, 'The Spaces in Between: Science, Ocean, Empire', *Isis* 105 (2014), pp. 338–51.
54 John Murray and Alphonse-François Renard, 'On the microscopic characters of volcanic ashes and cosmic dust, and their distribution in the deep sea deposits', *Proceedings of the Royal Society of Edinburgh* 12 (1884), pp. 474–95.
55 Martyn Low and Neal Evenhuis, 'Dates of publication of the Zoology parts of the *Report on the Scientific Results of the Voyage of H.M.S. Challenger During the Years 1873–76*', *Zootaxa* 3701 (2013), pp. 401–20, p. 402. For images of the site, see National Library of Scotland (NLS), Edinburgh, Neill & Company, printers, Edinburgh, ref. Dep. 196: 42 (photographs of plant, premises and presses, nineteenth to twentieth century).
56 *History of the Firm of Neill & Company, Ltd* (Edinburgh: Neill & Company, 1900).
57 Ibid.
58 NLS, Neill & Company, ref. Dep. 196: 35 (volume listing bills payable and receivable, 1875–97); 37 (wage list, 1895).
59 NLS, Neill & Company, ref. Dep. 196: 35, 37, 42
60 NLS, Neill & Company, ref. Dep. 196: 36 (correspondence relating to business premises, 1885–1900).
61 Examples of proof sheets from *Challenger Reports*, University of Edinburgh, HMS *Challenger* Papers, 1872–1876, Coll-46 at Gen. 28–31.
62 MCZ, Letter from John Murray to Alexander Agassiz, 22 February 1891, bAg 654.10.1.
63 Ailsa Allaby and Michael Allaby, 'Globigerina ooze', *A Dictionary of Earth Science*, 8 May 2018, https://www.encyclopedia.com (accessed 10 August 2022).
64 Ailsa Allaby and Michael Allaby, 'Red Clay', *A Dictionary of Earth Science*, 8 May 2018, www.encyclopedia.com (accessed 10 August 2022).
65 Ailsa Allaby and Michael Allaby, 'Diatom Ooze', *A Dictionary of Earth Science*, 8 May 2018, https://www.encyclopedia.com (accessed 10 August 2022).
66 'The "*Challenger*" report and the government', *The Pall Mall Gazette*, 18 April 1895.
67 The author viewed the *Challenger Report* series at the Stazione Zoologica in Naples and at the Museum of Comparative Zoology at Harvard University, Cambridge, Massachusetts.
68 For the expedition report as a particular style and its importance to ocean science, see Nyhart, 'Voyaging and the Scientific Expedition Report, 1800–1940', pp. 71–73, 79–80.

CONCLUSION

1 John Murray, *A Summary of the Scientific Results* [*Challenger Report*], 2 vols, (Edinburgh: HMSO, 1895), vol. 1, pt. 1, p. 2.
2 John Urry, *Mobilities* (Cambridge: Polity, 2007), p. 13.
3 Marjo K. Vierros, Autumn-Lynn Harrison, Matthew R. Sloat, Guillermo Ortuño Crespo, Jonathan W. Moore, Daniel C. Dunn, Yoshitaka Ota, Andrés M. Cisneros-Montemayor, George L. Shillinger, Trisha Kehaulani Watson, Hugh Govan, 'Considering Indigenous Peoples and local communities in governance of the global ocean commons', *Marine Policy* 119 (2020), pp. 1–13, p. 1. doi.org/10.1016/j.marpol.2020.104039 (accessed 6 September 2022).
4 Our Shared Seas, 'Threats: Habitats and Biodiversity Loss', https://oursharedseas.com/threats/threats-habitat-and-biodiversity/#:~:text=Human%20pressures%20are%20driving%20the,climate%20change%20and%20ocean%20acidification (accessed 5 September 2022).

FURTHER READING

RECOMMENDED WEBSITES

History of the Atlantic Cable & Undersea Communications
https://atlantic-cable.com/

History of Oceanography
Official website of the International Commission
of the History of Oceanography
https://oceansciencehistory.com/

HMS Challenger Collections
https://www.hmschallenger.net/

Sea Change: Celebrating the groundbreaking expedition
of HMS *Challenger*
The University of Edinburgh
https://exhibitions.ed.ac.uk/exhibitions/sea-change

All web addresses accurate at time of publication.

PRIMARY SOURCES

Most of the following published primary sources, as well as the 50 volumes of the *Challenger Report*, are available on Biodiversity Heritage Library (www.biodiversitylibrary.org)

Agassiz, Alexander, *Revision of the Echini, Illustrated Catalogue of the Museum of Comparative Zoölogy at Harvard College*, No. 7 (Cambridge, MA: Cambridge University Press, 1872–1874)

Agassiz, Alexander, *A Contribution to American Thalassography: Three Cruises of the United States Coast and Geodetic Survey Steamer 'Blake', in the Gulf of Mexico, in the Caribbean Sea, and along the Atlantic Coast of the United States, from 1877 to 1880*, 2 vols (Boston and New York: Houghton, Mifflin and Company, 1888)

Agassiz, George Russell (ed.), *Letters and Recollections of Alexander Agassiz: with a sketch of his life and work* (Boston: Houghton Mifflin Company, 1913)

Campbell, Lord George, *Log-Letters from the 'Challenger'* (London: Macmillan, 1877)

Maury, Matthew Fontaine, *The Physical Geography of the Sea* (London: Sampson, Low, Son & Co., 1855)

Moseley, Henry Nottidge, *Notes by a Naturalist: Observations Made During the Voyage of H.M.S. 'Challenger' Round the World in the Years 1872–1876* (London: John Murray, 1892)

Murray, John, and Hjort, Johan, *The Depths of the Ocean: a general account of the modern science of oceanography based largely on the scientific researches of the Norwegian steamer Michael Sars in the North Atlantic* (London: Macmillan, 1912)

Murray, John, and Renard, Alphonse-François, *Report on Deep-Sea Deposits Based on the Specimens Collected During the Voyage of H.M.S. Challenger in the Years 1872 to 1876 [Challenger Report]* (Edinburgh: HMSO, 1891)

Spry, William J.J., *The Cruise of Her Majesty's Ship 'Challenger'* (London: Sampson Low, Marston, Searle, and Rivington, 1876)

Swire, Herbert, *The Voyage of the Challenger* (London: Golden Cockerel Press, 1938)

Thomson, Charles Wyville, *The Depths of the Sea: an account of the general results of the dredging cruises of H.M.S.s 'Porcupine' and 'Lightning' during the summers of 1868, 1869, and 1870, under the scientific direction of Dr. Carpenter, F.R.S., J. Gwyn Jeffreys, F.R.S., and Dr. Wyville Thomson, F.R.S.* (London: Macmillan, 1873)

Thomson, Charles Wyville, *The Voyage of the 'Challenger': The Atlantic; a Preliminary Account of the General Results of the Exploring Voyage of H.M.S. 'Challenger' During the Year 1873 and the Early Part of the Year 1876*, 2 vols (London: Macmillan, 1877)

Thomson, Charles Wyville, *General Introduction to the Zoological Series of Reports, [Challenger Report: Zoology]* (Edinburgh: HMSO, 1880), vol. 1, pt. 1

Tizard, Thomas Henry, Moseley, Henry Nottidge, Buchanan, John Young, and Murray, John, *Narrative of the Cruise of H.M.S. Challenger, with a General Account of the Scientific Results of the Expedition, 2 vols [Challenger Report]* (Edinburgh: HMSO, 1885)

Wild, John James, *At Anchor: A Narrative of Experiences Afloat and Ashore During the Voyage of H.M.S. 'Challenger' from 1872 to 1876* (London and Belfast: M. Ward and Co., 1878)

SECONDARY SOURCES

Adler, Anthony, 'The ship as laboratory: making space for field science at sea', *Journal of the History of Biology* 47 (2014), pp. 333–62.

Adler, Anthony, *Neptune's Laboratory: Fantasy, Fear, and Science at Sea* (Cambridge, MA and London: Harvard University Press, 2019)

Anim-Addo, Anyaa,"A Wretched and Slave-Like Mode of Labor': Slavery, Emancipation and the Royal Mail Steam Packet Company's Coaling Stations', *Historical Geography* 39 (2011), pp. 65 –84

Beckert, Sven, *Empire of Cotton: A New History of Global Capitalism* (London: Penguin Random House, 2015)

John F. Beeler, *British Naval Policy in the Gladstone-Disraeli Era, 1866–1880* (Stanford, California: Stanford University Press, 1997)

Jeremy Black, 'The Victorian Maritime Empire in Its Global Context', in Miles Taylor (ed.), *The Victorian Empire and Britain's Maritime World, 1837–1901: The Sea and Global History* (New York: Palgrave Macmillan, 2013), pp. 167–88

Harold L. Burstyn, 'Pioneering in Large-scale Scientific Organisation: The Challenger Expedition and its Report. I. Launching the Expedition', *Proceedings of the Royal Society of Edinburgh, Section B: Biological Sciences* 72 (1972), pp. 47–61

Harold L. Burstyn, 'Science Pays Off: Sir John Murray and the Christmas Island Phosphate Industry, 1886–1914', *Social Studies of Science* 5 (1975), pp. 5–34

Eileen V. Brunton, *The Challenger Expedition, 1872–1876: A Visual Index* (London: Natural History Museum, 1994)

Rosamunde Codling, 'H.M.S. Challenger in the Antarctic: Pictures and Photographs from 1874', *Landscape Research* 22 (1997), pp.191–208

Margaret Cohen (ed.), *A cultural history of the sea*, 6 vols (London and New York: Bloomsbury, 2021)

Richard Corfield, *The Silent Landscape: The Scientific Voyage of HMS Challenger* (Washington, DC: Joseph Henry Press, 2003)

Lorraine Daston, 'Type Specimens and Scientific Memory', *Critical Inquiry* 31 (2004), pp. 153–182

Margaret Deacon, *Scientists and the Sea, 1650–1900: A Study of Marine Science* (Aldershot: Ashgate, 1997)

Margaret Deacon, Tony Rice and Colin Summerhayes (eds), *Understanding the Oceans: A Century of Ocean Exploration* (London and New York: UCL Press, 2001)

Sarah Dry, *Waters of the World: The Story of the Scientists Who Unraveled the Mysteries of Our Oceans, Atmosphere, and Ice Sheets and Made the Planet Whole* (Chicago: University of Chicago Press, 2019)

Erik Dücker, *News from an Inaccessible World: The History and Present Challenges of Deep-Sea Biology* (Enschede, Netherlands: Gildeprint, 2014)

Carolyn Fry, *Mapping the Oceans: Discovering the world beneath our seas* (London: Arcturus, 2020)

Aileen Fyfe, *Steam-Powered Knowledge: William Chambers and the Business of Publishing, 1820–1860* (Chicago: University of Chicago Press, 2012)

Daniel R. Headrick, *The Invisible Weapon: Telecommunications and International Politics, 1851–1945* (Oxford: Oxford University Press, 1991)

Henriet, Jean-Pierre, 'The Face of the Ocean: Alphonse-François Renard (1842–1903) and the Rise of Marine Geology', *Sartoniana* 23 (2010), pp. 47–80

Hood, Stephanie, 'Science, objectivity and photography in the nineteenth century: photographs from the voyage of HMS *Challenger* 1872–1876', MSc thesis, University College London, 2013.

Jeremy, John, *Cockatoo Island: Sydney's Historic Dockyard* (Sydney: University of New South Wales Press, 2005)

Jones, Erika Lynn, 'Making the Ocean Visible: Science and Mobility on the Challenger Expedition, 1872–1895', PhD thesis, University College London, 2019

Jones, Ian, and Jones, Joyce, *Oceanography in the Days of Sail* (Sydney: Hale & Iremonger, 1993)

Klein, Maury, *Union Pacific: The Reconfiguration: America's Greatest Railroad from 1869 to the Present* (Oxford: Oxford University Press, 2011)

Lambert, Andrew, 'Economic Power, Technological Advantage, and Imperial Strength: Britain as a Unique Global Power, 1860–1890', *International Journal of Maritime History* 5 (2006)

Lighman, Bernard (ed.) *Victorian Science in Context* (Chicago and London: University of Chicago Press, 1997)

Linklater, Eric, *The Voyage of the Challenger* (London: John Murray, 1972)

Livingstone, David N., *Putting Science in Its Place: Geographies of Scientific Knowledge* (Chicago and London: University of Chicago Press, 2003)

Livingstone, David N., and Withers, Charles W. J. (eds), *Geographies of Nineteenth-Century Science* (Chicago and London: University of Chicago Press, 2011)

Low, Martyn and Evenhuis, Neal, 'Dates of publication of the Zoology parts of the Report on the Scientific Results of the Voyage of H.M.S. Challenger During the Years 1873–76', *Zootaxa* 3701 (2013), pp. 401–20

MacDougall, Philip, *Chatham Dockyard: The Rise and Fall of a Military Industrial Complex* (Stroud, Gloucestershire: The History Press, 2012)

Marsden, Ben, and Smith, Crosbie, *Engineering Empires: A cultural history of technology in nineteenth-century Britain* (Basingstoke: Palgrave Macmillan, 2005)

McConnell, Anita, *No Sea Too Deep: The History of Oceanographic Instruments* (Bristol: Hilger, 1982)

Millar, Sarah Louise, 'Science at Sea: Soundings and Instrumental Knowledge in British Polar Expedition Narratives, *c.*1818–1848', *Journal of Historical Geography* 42 (2013), pp. 77–87

Mills, Eric L., *Biological Oceanography: an Early History, 1870–1960* (Toronto: University of Toronto Press, 2012)

Nyhart, Lynn, 'Voyaging and the scientific expedition report, 1800–1940', in Apple, Rima D., Downey, Gregory J. and Vaughn, Stephen L. (eds), *Science in Print: Essays on the History of Science and the Culture of Print* (Madison, WI: University of Wisconsin Press, 2012), pp. 65–86.

Paton, Lucy Allen, *Elizabeth Cary Agassiz: A Biography* (Boston: Houghton Mifflin Company, 1919)

Pearson, Philip, *A Challenger's Song* (London: Austin Macauley Publishers, 2021)

Philbrick, Nathaniel, 'The United States Exploring Expedition, 1838–1842', Smithsonian Libraries, 2004, https://www.sil.si.edu/DigitalCollections/usexex/learn/Philbrick.htm

Raj, Kapil, 'Beyond Postcolonialism...and Postpositivism: Circulation and the Global History of Science', *Isis* 104 (2003), pp. 337–47

Rehbock, Philip F. (ed.), *At Sea with the Scientifics: The Challenger Letters of Joseph Matkin* (Honolulu: University of Hawaii Press, 1992)

Reidy, Michael S., *The Tides of History: Ocean Science and Her Majesty's Navy* (Chicago: University of Chicago Press, 2008)

Reidy, Michael S., Kroll, Gary R., and Conway, Erik M., *Exploration and Science: Social Impact and Interaction* (Oxford: ABC-CLIO, 2007)

Reidy, Michael S., and Rozwadowski, Helen M., 'The Spaces in Between: Science, Ocean, Empire', *Isis* 105(2014), pp. 338–51

Rozwadowski, Helen M., *Fathoming the Ocean: The Discovery and Exploration of the Deep Sea* (Cambridge, MA: The Belknap Press, 2005)

Rozwadowski, Helen M., *Vast Expanses: A History of the Oceans* (London: Reaktion Books, 2018)

Ryan, James R., *Photography and Exploration* (London: Reaktion Books, 2013)

Ryan, James R., *Picturing Empire: Photography and the Visualization of the British Empire* (London: Reaktion Books, 1997)

Schlee, Susan, *The Edge of an Unfamiliar World: A History of Oceanography* (New York: Dutton, 1973)

Schurman, D. M., *The Education of a Navy: The Development of British Naval Strategic Thought, 1867–1914* (London: Cassell and Co., 1965)

Secord, James A., *Victorian Sensation: The Extraordinary Publication, Reception, and Secret Authorship of Vestiges of the Natural History of Creation* (Chicago: University of Chicago Press, 2000)

Sorrenson, Richard, 'The Ship as a Scientific Instrument in the Eighteenth Century', *Osiris* 11 (1996), pp. 221–36

Spence, Daniel Owen, *A History of the Royal Navy: Empire and Imperialism* (London and New York: I. B. Tauris, 2015)

Stein, Glenn M., 'The Challenger Medal Roll (1895)', 2007, http://www.19thcenturyscience.org/HMSC/Chall-Medal/ChallengerMedal.html

Wenzlhuemer, Roland, *Connecting the Nineteenth-Century World: The Telegraph and Globalization* (Cambridge: Cambridge University Press, 2012)

Winsor, Mary P., *Reading the Shape of Nature: Comparative Zoology at the Agassiz Museum* (Chicago: University of Chicago Press, 1991)

Zuroski, Emma, 'Situating the local in a global expedition: HMS Challenger Expedition in New Zealand, 1874', *Journal of the Royal Society of New Zealand* 47 (2017), pp. 107–11.

ACKNOWLEDGEMENTS

Many individuals aided in the writing and production this book, and I am happy to acknowledge them here. I first began investigating the topic of the *Challenger* Expedition in 2014 as the focus of my doctoral thesis, a UK Arts and Humanities Research Council collaborative doctoral award undertaken with the Department of Science and Technology Studies at University College London and the National Maritime Museum, Greenwich. Many thanks are due to these institutions and my PhD supervisors, Simon Werrett and Richard Dunn.

I received kind assistance from numerous people at the following institutions: University of Edinburgh Centre for Research Collections, Science Museum, Natural History Museum, UK Hydrographic Office, Royal Geographical Society and the Caird Library and Archives at the National Maritime Museum. Furthermore, I am grateful for my time as a Research Fellow at the Smithsonian Institution Archives, Washington, DC. Pamela M. Henson aided my understanding of nineteenth-century American science and Dave Pawson, Curator of Echinoderms at the National Museum of Natural History, sparked my interest in Alexander Agassiz and supported my subsequent archival visit to the Ernst Mayr Library at the Museum of Comparative Zoology at Harvard University.

I owe an immense debt of gratitude to those who helped transform the academic thesis into an engaging narrative for a wider audience. National Maritime Museum curators Robert Blyth, Louise Devoy and Jeremy Michell reviewed draft chapters, as did historians of ocean science, medicine and technology Penelope Hardy and Sam Robinson — all significantly improved the book's final form.

Exploring *Challenger*'s collection of photographs and its representation of the people of Te-Moana-Nui-A-Kiwa (Pacific Ocean) is a new area of historical research and I am grateful for the perspective and knowledge of the Tangata Moana Advisory Board. One of its members, Katrina Igglesden, made important suggestions regarding the text, as did Aaron Jaffer and Rebecca Martin.

This title would not have been possible without the dedication and tireless work of Kathleen Bloomfield, who oversaw the book's production at the National Maritime Museum. As part of her multifaceted duties, Louise Jarrold sourced the array of beautiful illustrations found throughout.

Finally, I am thankful for the support and encouragement of my friends, parents Chris and Sam, my in-laws Mary and Ted and, especially, my loving husband Phil.

PICTURE CREDITS

The publisher would like to thank the copyright holders for granting permission to reproduce the images illustrated. Every attempt has been made to trace accurate ownership of copyrighted images in this book. Any errors or omissions will be corrected in subsequent editions provided notification is sent to the publisher.

Unless otherwise stated, images are © National Maritime Museum, Greenwich, London. Object IDs for objects from the collection of Royal Museums Greenwich are included below. Further information about the Museum and its collection can be found at rmg.co.uk.

ON THE FRONT COVER

From Charles Wyville Thomson, *The Depths of the Sea: an account of the general results of the dredging cruises of H.M.S.s 'Porcupine' and 'Lightning' during the summers of 1868, 1869, and 1870, under the scientific direction of Dr. Carpenter, F.R.S, J. Gwyn Jeffreys, F.R.S, and Dr. Wyville Thomson, F.R.S* (London: Macmillan and Co., 1874), Figure 5. PBG3432

ON THE BACK COVER

From Thomson, *The Depths of the Sea*, Figure 42. PBG3432

INSIDE COVER

From Thomson, *The Depths of the Sea*, Figures 4, 30, 72, 78 and 79. PBG3432

From Alexander Agassiz, *Report on the Echinoidea Dredged by H.M.S. Challenger, During the Years 1873–1876' [Challenger Report: Zoology]*, vol. 3, pt. 9, Plate IV. The University of Edinburgh – CC BY 3.0 (https://creativecommons.org/licenses/by/3.0/) (details, colour adapted)

pp. 4–5 ALB0175
pp. 12–13 PAD6215
p. 14 PAF0588
p. 16 From John Ross, *A Voyage of Discovery, Made under the Orders of the Admiralty, in His Majesty's Ships Isabella and Alexander, for the purpose of exploring Baffin's Bay, and Inquiring into the Possibility of a Northwest Passage* (London: John Murray, 1819). PBC4991
p. 17 © Science Museum Group
p. 19 From Thomson, *The Depths of the Sea*, Figure 1. PBG3432
p. 20 PAD6090
p. 21 BHC2981 © National Maritime Museum, Greenwich, London. Caird Fund.
p. 22 From Sir John Ross, *Narrative of a Second Voyage in Search of a North-West Passage*, Natural History appendix (London: A. & W. Galignani, 1835), Plate B. © The Royal Society
p. 24 PAF0588
p. 25 From Charles Darwin, *On the Structure and Distribution of Coral Reefs: also geological observations on the volcanic islands and parts of South America* (London: Ward, Lock and Co., 1890). PBB4912
p. 26 From Edward Forbes and Robert Godwin-Austen (eds), *The Natural History of the European Seas* (London: J. Van Voorst, 1859). Steve Nicklas, National Ocean Service/ National Oceanic and Atmospheric Administration
p. 28 From Joseph-Fortuné-Théodose Eydoux and Louis-François-Auguste Souleyet, *Voyage autour du monde exécuté pendant les années 1836 et 1837 sur la corvette la Bonite, commandée par M. Vaillant. Histoire naturelle. Zoologie. Atlas* (Paris: A. Bertrand, 1851), Mollusques, Plate 23. Bibliothèque nationale de France
p. 29 (top) From Joseph-Fortuné-Théodose Eydoux and Louis-François-Auguste Souleyet, *Voyage autour du monde exécuté pendant les années 1836 et 1837 sur la corvette la Bonite, commandée par M. Vaillant. Histoire naturelle. Zoologie. Atlas* (Paris: A. Bertrand, 1841–1852), Poissons, Plate 2. © Heritage Collections, Auckland Libraries
p. 29 (bottom) NAV1003
p. 32 From Thomson, *The Depths of the Sea*, Figure 39. PBG3432
pp. 34–5 From Matthew Fontaine Maury, *The Physical Geography of the Sea, and its Meteorology* (London: Sampson Low, Son & Co., 1861), Plate XI. PBB5002
p. 36 ZBA0001.5
p. 37 PAD6212
p. 39 The Metropolitan Museum of Art
p. 40 From Thomson, *The Depths of the Sea*, Figure 4. PBG3432
p. 41 From Thomson, *The Depths of the Sea*, Figure 72. PBG3432
p. 42 © National Maritime Museum, Greenwich, London. Reproduced with kind permission of the artist's son and grandsons. SLR0709
p. 43 From Thomson, *The Depths of the Sea*, Plate I. PBG3432
p. 44 From Thomson, *The Depths of the Sea*, Figure 5. PBG3432
p. 43 From Thomson, *The Depths of the Sea*, Figure 51. PBG3432
p. 47 © NIWA/ Rob Stewart
p. 48 ALB0174
p. 50 PAD6215
p. 54–5 ALB0175
p. 58 (both) Thomas Henry Tizard, Henry Nottidge Moseley, John Young Buchanan, and John Murray, *Narrative of the Cruise of H.M.S. Challenger, with a General Account of the Scientific Results of the Expedition [Challenger Report]*, 2 vols (Edinburgh: HMSO, 1885), vol. 1. Image from the Biodiversity Heritage Library. Contributed by University of Toronto – Gerstein Science Information Centre. | www.biodiversitylibrary.org
p. 59 From John Murray and Alphonse-François Renard, *Report on Deep-Sea Deposits Based on the Specimens Collected During the Voyage of H.M.S. Challenger in the Years 1872 to 1876 [Challenger Report]* (Edinburgh: HMSO, 1891), Figure 17. PBP7849 © Crown copyright. Photo © National Maritime Museum, Greenwich, London
pp. 60–1 NPA8446 © Crown copyright. Photo © National Maritime Museum, Greenwich, London
p. 62 NPA8446 (detail) © Crown copyright. Photo © National Maritime Museum, Greenwich, London
p. 63 NPA8446 (detail) © Crown copyright. Photo © National Maritime Museum, Greenwich, London
p. 64 From Tizard *et al.*, *Narrative of the Cruise*, vol. 1, Figure 5. The University of Edinburgh – CC BY 3.0 (https://creativecommons.org/licenses/by/3.0/) (detail)
pp. 66–7 NPA8447 © Crown copyright. Photo © National Maritime Museum, Greenwich, London
p. 68 From Tizard et *al.*, *Narrative of the Cruise*, vol. 1. Image from the Biodiversity Heritage Library. Contributed by Natural History Museum Library, London. | www.biodiversitylibrary.org
p. 70 ALB0174
p. 72 ALB0174
p. 73 ALB0174
p. 75 ALB0175
p. 77 Photo supplied by Martin Bosher
p. 78 From Tizard et al., *Narrative of the Cruise*, vol. 1, Figure 14. The University of Edinburgh – CC BY 3.0 (https://creativecommons.org/licenses/by/3.0/) (detail)
p. 80 From Tizard *et al.*, *Narrative of the Cruise*, vol. 1. Image from the Biodiversity Heritage Library. Contributed by Natural

History Museum Library, London. | www.biodiversitylibrary.org
p. 81 PBD8264 © National Maritime Museum, Greenwich, London, Macpherson Collection
p. 82 © Science Museum Group
p. 83 (top) Image © National Museums Scotland
p. 83 (bottom) From Thomson, *The Depths of the Sea*, Figure 42. PBG3432
p. 83 N5395
p. 86 (top) Image © National Museums Scotland
p. 86 (bottom) From Tizard *et al.*, *Narrative of the Cruise*, vol. 1, Figure 14. The University of Edinburgh – CC BY 3.0 (https://creativecommons.org/licenses/by/3.0/) (detail)
p. 90 From Tizard *et al.*, *Narrative of the Cruise*, vol. 1, Figure 16. The University of Edinburgh – CC BY 3.0 (https://creativecommons.org/licenses/by/3.0/) (detail)
p. 91 From Tizard *et al.*, *Narrative of the Cruise*, vol. 1, Figure 15. Image from the Biodiversity Heritage Library. Contributed by University of Toronto – Gerstein Science Information Centre. | www.biodiversitylibrary.org
p. 92 ALB0174
p. 97 © Special Collections & Archives, University of California San Diego
p. 98 © The Trustees of the Natural History Museum, London
p. 99 From Murray and Renard, *Report on Deep-Sea Deposits*, Diagram 15. PBP7849, © Crown copyright. Photo © National Maritime Museum, Greenwich, London
p. 100 Image © National Museums Scotland
p. 101 From *H.M.S Challenger, Reports of Captain G.S. Nares, RN: with abstract of soundings & diagrams of ocean temperature in North and South Atlantic Oceans, 1873* (London: HMSO, 1873). JOD/15/2/1 © Crown copyright. National Maritime Museum, Greenwich, London
p. 102 From Edgar A. Smith, *Report on the Lamellibranchiata Collected by H.M.S Challenger During the Years 1873-76 [Challenger Report]* (Edinburgh: HMSO, 1885), Plate XV. The University of Edinburgh – CC BY 3.0 (https://creativecommons.org/licenses/by/3.0/) (detail)
p. 104 From Frederick Whymper, *The Sea: Its Stirring Story of Adventure, Peril & Heroism* (London: Cassell, 1887). British Library, London, UK © British Library Board. All Rights Reserved / Bridgeman Images
p. 105 From Smith, *Report on the Lamellibranchiata*, Plate XV. The University of Edinburgh – CC BY 3.0 (https://creativecommons.org/licenses/by/3.0/) (detail)
p. 106 From Thomson, *The Depths of the Sea*, Figures 78 and 79. PBG3432
p. 109 From John James Wild, *At Anchor: A Narrative of Experiences Afloat and Ashore During the Voyage of H.M.S. 'Challenger' from 1872 to 1876* (London and Belfast: M. Ward and Co., 1878). PBD4528
p. 110 From Tizard *et al.*, *Narrative of the Cruise*, vol. 1. Image from the Biodiversity Heritage Library. Contributed by University of Toronto – Gerstein Science Information Centre. | www.biodiversitylibrary.org
p. 111 From Wild, *At Anchor: A Narrative of Experiences Afloat and Ashore During the Voyage of H.M.S. 'Challenger' from 1872 to 1876*. PBD4528
p. 112 From Murray and Renard, *Report on Deep-Sea Deposits*, Chart 21. PBP7849 © Crown copyright. Photo © National Maritime Museum, Greenwich, London
p. 113 ALB0175
p. 114 ALB0175
p. 116 ALB0174
p. 117 From Tizard *et al.*, *Narrative of the Cruise*, vol. 1. Image from the Biodiversity Heritage Library. Contributed by University of Toronto – Gerstein Science Information Centre. | www.biodiversitylibrary.org
pp. 118–19 From Tizard *et al.*, *Narrative of the Cruise*, vol. 1. The University of Edinburgh – CC BY 3.0 (https://creativecommons.org/licenses/by/3.0/) (detail)
p. 120 ALB0175
p. 124 © The Trustees of the Natural History Museum, London
p. 126 ALB0174
p. 128 ALB0175
p. 129 © National Portrait Gallery, London
p. 130 © Imperial War Museum (Q 69834)
pp. 132–3 NPA8446 (detail) © Crown copyright. National Maritime Museum, Greenwich, London
p. 134 From Sir William de Wiveleslie Abney, *A Treatise on Photography* (London: Longmans & Co., 1878). British Library, London, UK / © British Library Board. All Rights Reserved / Bridgeman Images
p. 135 NPA8446 (detail) © Crown copyright. National Maritime Museum, Greenwich, London
p. 136 ALB0174
pp. 138–9 ALB0174
p. 140 ALB0174
p. 141 From Tizard *et al.*, *Narrative of the Cruise*, vol. 1, Figure 59. Image from the Biodiversity Heritage Library. Contributed by University of Toronto – Gerstein Science Information Centre. | www.biodiversitylibrary.org
pp. 142–3 ALB0174
p. 144 ALB0174
p. 145 From Lady Charlotte Anna Lefroy, *Flowers of Bermuda 1871–1877* © Bermuda National Trust Collection
p. 146 From Tizard *et al.*, *Narrative of the Cruise*, vol. 1, Plate IV. Image from the Biodiversity Heritage Library. Contributed by University of Toronto – Gerstein Science Information Centre. | www.biodiversitylibrary.org
p. 147 © Illustrated London News Ltd/Mary Evans
p. 149 ALB0175
p. 150 ALB0175
p. 151 ALB0175
p. 152 State Library Victoria
p. 153 (top) ALB0175
p. 153 (bottom) From Tizard *et al.*, *Narrative of the Cruise*, vol. 1, Sheet 28. Image from the Biodiversity Heritage Library. Contributed by University of Toronto – Gerstein Science Information Centre. | www.biodiversitylibrary.org
p. 154 ALB0175
p. 155 ALB0175
p. 156 ALB0175
p. 157 ALB0175
p. 158 ALB0175
p. 159 ALB0174
p. 160 From Wild, *At Anchor: A Narrative of Experiences Afloat and Ashore During the Voyage of H.M.S. 'Challenger' from 1872 to 1876*. PBD4528
p. 162 ALB0174
p. 163 (both) ALB0174
p. 164 ALB0174
p. 165 ALB0174
p. 166 ALB0176
p. 168 From Alexander Agassiz, *Revision of the Echini, Illustrated Catalogue of the Museum of Comparative Zoölogy at Harvard College*, No. 7 (Cambridge, MA: Cambridge University Press, 1872–1874), Plates, Plate II. Image from the Biodiversity Heritage Library. Contributed by Harvard University, Museum of Comparative Zoology, Ernst Mayr Library. | www.biodiversitylibrary.org
p. 170 From Agassiz, *Report on the Echinoidea Dredged by H.M.S. Challenger, During the Years 1873–1876*, Plate IV. The University of Edinburgh – CC BY 3.0 (https://creativecommons.org/licenses/by/3.0/) (detail)
p. 172 From the Ernst Mayr Library and Archives of the Museum of Comparative Zoology, Harvard University.
p. 174 (both) From George Russell Agassiz (ed.), *Letters and Recollections of Alexander*

Agassiz: with a sketch of his life and work (Boston: Houghton Mifflin and Company, 1913)
p. 176 From Agassiz (ed.), *Letters and Recollections of Alexander Agassiz*
p. 178 From Agassiz, *Revision of the Echini*, Plates, Plate II. Image from the Biodiversity Heritage Library. Contributed by Harvard University, Museum of Comparative Zoology, Ernst Mayr Library. | www.biodiversitylibrary.org
p. 179 From Alexander Agassiz, *Revision of the Echini*, Plates, Plate 1f. © The Royal Society
p. 183 Roosevelt 560.12-028. Houghton Library, Harvard University
p. 185 From Agassiz (ed.), *Letters and Recollections of Alexander Agassiz*
p. 188 From Ernst Haeckel, *Report on the Radiolaria Collected by H.M.S. Challenger During the Years 1873–1876* [Challenger Report: Zoology] (Edinburgh: HMSO, 1887), vol. 18, pt. 40, Plate 60. The University of Edinburgh – CC BY 3.0 (https://creativecommons.org/licenses/by/3.0/) (detail)
p. 189 From Georg Ossian Sars, *Report on the Schizopoda Collected by H.M.S. Challenger During the Years 1873–1876* [Challenger Report: Zoology] (Edinburgh: HMSO, 1885), vol. 13, pt. 37, Plate XXXVI. British Library, London, UK / © British Library Board. All Rights Reserved / Bridgeman Images
p. 191 © NIWA / Sadie Mills
p. 192 From Thomas Henry Huxley and Paul Pelseneer, *Report on the Specimen of the Genus Spirula Collected by H.M.S. Challenger [Challenger Report: Zoology]* (Edinburgh: HMSO, 1895), Appendix, Plate III. Image from the Biodiversity Heritage Library. Contributed by Natural History Museum, London. | www.biodiversitylibrary.org
p. 194 Library of Congress, Manuscript Division, Brady-Handy Collection
p. 196 From Murray and Renard, *Report on Deep-Sea Deposits*, Plate XVII. PBP7849, © Crown copyright. Photo © National Maritime Museum, Greenwich, London
p. 199 From Murray and Renard, *Report on Deep-Sea Deposits*, Plate XVIII. PBP7849, © Crown copyright. Photo © National Maritime Museum, Greenwich, London
p. 201 © Natural History Museum, London / Bridgeman Images
pp. 202–3 From Lord George Campbell, *Log-Letters from the 'Challenger'* (London: Macmillan, 1877). PBB3924
p. 206 From Murray and Renard, *Report on Deep-Sea Deposits*, Plate XXIII. The University of Edinburgh – CC BY 3.0 (https://creativecommons.org/licenses/by/3.0/) (detail).
p. 207 © Natural History Museum, London / Bridgeman Images
p. 208 From Ferdinand Zirkel, *Microscopical Petrography*, vol. 6, in Clarence King, Geologist-in-Charge, *United States Report of the Geological Exploration of the Fortieth Parallel* (Washington, DC: Government Printing Office, 1876), Plate XI. British Library, London, UK / © British Library Board. All Rights Reserved / Bridgeman Images
p. 209 From Charles de la Vallée Poussin and Alphonse-François Renard, *Mémoire sur les caractères minéralogiques et stratigraphiques des roches dites plutoniennes de la Belgique et de l'Ardenne française* (Brussels: Académie Royale de Belgique, 1876), Plate III. British Library, London, UK / © British Library Board. All Rights Reserved / Bridgeman Images
p. 212 From Murray and Renard, *Report on Deep-Sea Deposits*, Plate XXIII. The University of Edinburgh – CC BY 3.0 (https://creativecommons.org/licenses/by/3.0/) (detail)
p. 213 From Murray and Renard, *Report on Deep-Sea Deposits*, Plate XVII. PBP7849, © Crown copyright. Photo © National Maritime Museum, Greenwich, London
p. 216 From Murray and Renard, *Report on Deep-Sea Deposits*, Plate XII. PBP7849, © Crown copyright. Photo © National Maritime Museum, Greenwich, London
p. 217 From Murray and Renard, *Report on Deep-Sea Deposits*, Plate XII. PBP7849, © Crown copyright. Photo © National Maritime Museum, Greenwich, London
p. 219 From the collection of Neill & Company deposited at the National Library of Scotland.
pp. 220–1 From Murray and Renard, *Report on Deep-Sea Deposits*, Chart I. The University of Edinburgh – CC BY 3.0 (https://creativecommons.org/licenses/by/3.0/) (detail)
pp. 228–9 From William J.J. Spry, *The Cruise of Her Majesty's Ship 'Challenger'* (London: Sampson Low, Marston, Searle and Rivington, 1876). PBB3932

INDEX

First published in 2022 by Royal Museums Greenwich,
Park Row, Greenwich, London, SE10 9NF

ISBN: 978-1-9063679-7-8

At the heart of the UNESCO World Heritage Site of Maritime Greenwich are the four world-class attractions of Royal Museums Greenwich — the National Maritime Museum, the Royal Observatory, the Queen's House and *Cutty Sark*.

rmg.co.uk

A CIP catalogue record for this book is available from the British Library.

Design by Rita Peres Pereira

Printed and bound in Slovenia by DZS Grafik